||||||||

||||||||

Acid Rain and Friendly Neighbors
The Policy Dispute between Canada
and the United States

||||||||

''''''''

''''''''

Acid Rain and Friendly Neighbors

The Policy Dispute between Canada and the United States

''''''''

Revised edition

Edited by Jurgen Schmandt, Judith Clarkson,

and Hilliard Roderick

Library of Congress Cataloging-in-Publication Data

Acid rain and friendly neighbors: the policy dispute between Canada
and the United States / edited by Jurgen Schmandt, Hilliard
Roderick, and Judith Clarkson.
p. cm. —(Duke Press policy studies)
Rev. ed. of: Acid rain and friendly neighbors / edited by Jurgen
Schmandt and Hilliard Roderick. 1985.
Bibliography: p. Includes index. |ISBN 0-8223-0870-3
1. Acid rain—Environmental aspects—United States. 2. Acid rain—
Environmental aspects—Canada. 3. Acid rain—Environmental
aspects—Government policy—United States. 4. Acid rain—
Environmental aspects—Government policy—Canada. I. Schmandt,
Jurgen. II. Roderick, Hilliard. III. Clarkson, Judith.
IV. Series.
TD196.A25A2814   1988
363.7'386—dc19   88-25605

In memory of

Hilliard Roderick

1923–1986

Contents

viii   Contents

‖‖‖‖‖

‖‖‖‖‖

Tables and Figures

‖‖‖‖‖

Tables

Figures

‖‖‖‖‖‖

‖‖‖‖‖‖

Preface

‖‖‖‖‖‖

This book presents the work of a policy research project on U.S.-Canadian relations as they pertain to the issue of acid rain. The project team consisted of fifteen graduate students and two faculty members—a public policy specialist and a physicist, both with administrative experience in national and international agencies working on environmental problems. (For a list of the team members, see the list of contributors at the end of this book.)

The project sought answers to three questions:

1. What is the extent of agreement and disagreement between Canada and the United States, as evidenced in published documents, about the nature of acid rain and its environmental effects?
2. What domestic policy developments are under way in Canada and the United States aimed at controlling acid rain, and what additional measures would be helpful?
3. What joint measures have the two countries taken to resolve the issue and, in light of the past record of resolving environmental disputes between the two countries, what additional bilateral measures can reasonably be expected?

Much has been said and written in recent years about the scientific and economic aspects of acid rain. Less analytical attention has been devoted to the policy side of the issue. Technical information is important because it alone can provide the basis for informed decision making on acid rain. Many questions of fact and interpretation remain unanswered, and improved knowledge is clearly needed. But policy development must proceed while scientific research continues. This volume concentrates on the policy questions raised by acid rain.

Because no acid rain policy is yet in place in the United States, much of the discussion is devoted to analysis of substantive and institutional factors that will determine an eventual solution.

We do not attempt to cover the whole range of policy issues. Formal negotiations between Canada and the United States have been under way since 1978. Several domestic policy measures to control acid rain have been taken or are under consideration, particularly in Canada. We provide background information on these matters, but our principal goal is to explore more general and varied ways in which Canada and the United States, jointly and individually, can search for answers to the acid rain problem.

The project was planned in 1981–82, the research was conducted in 1982–83, and the first edition of the volume was prepared for publication in 1983–84. The group used a common research design. Team members worked closely with the project directors, meeting weekly to discuss issues and progress or to hear from visiting experts. Sources of information included scholarly materials, government and industry documents, and interviews. An attempt was made throughout the study to represent, as objectively as possible, Canadian as well as U.S. viewpoints. An important element was a conference of U.S. and Canadian policymakers, analysts, and industry representatives that was hosted by the project in April 1983. The conference produced an intensive policy dialogue and clarified the areas of agreement and disagreement between the interests represented. Perhaps the most surprising thing we learned was that dialogues focusing on policy issues, rather than on scientific questions, occur infrequently, and some of the participants were meeting each other for the first time. The opportunity to meet occasionally for informal discussion is important in generating the appropriate atmosphere for a formal agreement. It is especially important in the absence of such an agreement, so that national policies are developed with, at least, sensitivity toward the views of the other parties.

In the spring of 1988 this volume was extensively revised and updated in preparation for publication of the second edition. For some chapters, such as those on bilateral actions or on domestic policy developments in Canada and the United States, this has resulted in the addition of new material to cover the events of the last three years. This information is now contained in chapters 3, 4, 6, and 7. The chapter on the role of interest groups (now chapter 5) has been extensively revised in order to explore in more depth the issues that are currently framing the debate. The two chapters on the International Joint Commission and its role in developing the Great Lakes Water

Quality Agreements have been consolidated into one chapter. The most difficult chapter to revise was chapter 2, in which a great number of U.S. and Canadian scientific assessments were compared. Because we did not have the resources to evaluate all of the new material in the comprehensive manner originally used, we chose to leave the chapter essentially unchanged, and added a section on the three most significant reports that have appeared since the original research was completed. We believe, however, that our conclusions on the status of the scientific assessments would not change if more recent documents were added to the analysis.

Funding for the project was provided by the Tom Slick Fund for World Peace. Support for the conference was provided by the Canadian embassy in Washington, D.C. Other support, including funding for the revision, was contributed by the Lyndon Baines Johnson School of Public Affairs and the University of Texas at Austin. Although the project received support from both Canadian and U.S. sources, we have been entirely independent in our work. Opinions expressed here are ours alone. Special thanks are due to our visitors, interviewees, and readers. Among team members we thank Andrew Morriss and Paul Kinscherff, who excelled as editorial assistants; Tom Albin and Kim DeRidder, who carried particularly heavy research and writing loads; and Susan Roush, who did everything, and more, that a project secretary does.

Acid rain is a difficult policy issue that clouds the relations between the two North American neighbors. We hope that this volume will be of help to the various groups and interests that ultimately will have to resolve the issue. As teachers of public policy, we have tried to give our students an opportunity to learn more about one of the new policy issues with which they will have to deal in their professional careers: problems created by man as unintended, dangerous byproducts of his very success in developing technology and industry. We believe that the need for skills and attributes to deal with such issues calls for a new approach to public service training. What is needed is an approach using methods analogous to the teaching hospital in medical education: hands-on experience under the guidance of teachers capable of dealing with practical issues in the light of recent advances in theory. The year-long policy research project, as it has been developed at the LBJ School since 1971, is a step in this direction. It provides faculty members and graduate students with an opportunity to research a policy issue, analyze and write up their results, and formulate policy recommendations. By now well over one hundred projects have been completed. We are pleased to present the product of one of them in this form.

"""""""

"""""""

Part One
The Search for a Bilateral Agreement

"""""""

Introduction to Part One

Acid rain, like many other environmental and health issues, presents the policymaker with an inherent difficulty. On one hand, without the benefit of scientific information, policy-making would be entirely in the dark—the nature of acid rain and its effects on lakes, forests, and man would be unknown. On the other hand, scientists point out that much remains unknown about the nature and effects of acid rain. Should policymakers wait for more complete information before taking action? If so, how complete?

We complain about the imperfect state of knowledge concerning acid rain, but our predicament would be immeasurably more serious, and without hope of eventual resolution, were it not for the scientific evidence that we do have. We would observe fish or forests dying but would have no idea why this was happening or where we might start looking for relief. This is how people must have felt about the plague in the Middle Ages or the dust bowl in more recent times. They faced a mysterious natural or God-sent disaster, and all man could do was wait for the end of the predicament or pack and go elsewhere. The knowledge we have about acid rain, although limited and at times controversial, makes for a different kind of situation. We have a basic understanding of the problem and its causes, a number of plausible working hypotheses about transformation and transport of pollutants, and a strategy for finding additional answers. Policymakers, using the evidence offered by scientific research, can engage in rational discourse about the problem, identify ways to control it, and test the efficacy and efficiency of alternative solutions.

Scientific information, obviously, is only part of the total picture, and other considerations—economic, social, and political—must be taken into account. It would be quite wrong to expect that science, once it told us what acid rain is and how it works, would determine the policy response. At most, it will help the policymaker to assess the seriousness of the problem and to distinguish between technically workable and impractical solutions. The policymaker, therefore, depends on scientific information; but final decisions will require judgments that only he can make.

These points are used in chapter 1 to develop the argument that acid rain is different in nature and effects from earlier forms of environmental pollution. We suggest that existing or proposed policy measures to control acid rain can alleviate the problem but will not be sufficient to deal with the full range of documented or suspected effects. The elements of a broad-based policy response are identified.

Scientific research on acid rain is conducted by many groups and in many places here, in Canada, and in other parts of the world. Research results are not immediately usable for policymaking. They first need to be summarized and interpreted—a task that requires scientific competence and the ability to present research results in nontechnical language. Scientific assessments are generally prepared by groups of experts and are later reviewed for accuracy and completeness by other groups of informed specialists. Sponsorship of assessments varies widely and may include government agencies, independent research organizations, industry, and public interest groups.

By now a variety of assessments have been prepared for use in policy discussions of acid rain. These have defined the important questions, summarized available knowledge, and identified as yet little-understood issues. Chapter 2 is designed to "assess the assessments" and examine their conclusions. How much agreement and disagreement is there among the various assessments? Is the range of disagreement as broad as the public debate on acid rain seems to suggest? Where there is disagreement among experts, what is it about: fundamental or secondary issues, questions of theory or specific findings, data, or interpretations? Is disagreement in the form of different conclusions drawn from the same evidence, or is the evidence itself controversial? To what extent are conclusions presented in the assessments a reflection of institutional sponsorship?

Examining more than a dozen assessments prepared from 1979 to 1983, and reviewing three later reports, chapter 2 finds that there is essential agreement among assessments on the most important issues:

There has been a historical increase in precursor emissions in North America.

Anthropogenic sources are more important in industrialized and urbanized regions than natural emissions of pollutants.

Long-range transport of acid and precursors occurs.

It is not (yet) possible to link specific sources with specific damages.

Acid deposition is likely to increase in the future.

Acid deposition is damaging some of the aquatic ecosystems of North America.

Damage is associated with reduced pH levels and the presence of toxic metals leached into surface water.

Acid shock due to springtime snowmelts results in greater damage than gradual acidification.

There is less agreement on the effects of acid deposition on terrestrial ecosystems. Some reports see the effects as serious, others do not. There is also disagreement on whether liming is an appropriate response to acidification. The nature of the disagreement in the two cases is different. For terrestrial effects, not enough is known to draw final conclusions, even though much research into the question has been conducted. On liming, most U.S. government reports see it as sometimes useful, while independent and Canadian reports state that it is not a promising response. Most documents, surprisingly, do not address the question of reversibility of effects. Those that do conclude that effects are not fully reversible.

The difference in U.S. and Canadian views on liming represents one of the few cases where experts from the two countries differ in their interpretation of the available scientific evidence. The institutions sponsoring the reports—government, industry, or independent organizations—seem to shade the assessment results more than might be expected just from national differences. This is particularly true with regard to the overall judgment on whether the effects of acid deposition are serious. Differences among institutional sponsors are less pronounced with respect to statements about the physical and chemical properties of acid rain.

Chapter 3 provides information about the diplomatic negotiations between Canada and the United States. Negotiations have been under way since 1978 and so far have not yielded an agreement. It is well known that the Reagan administration did not feel much urgency to advance the negotiations. Chapter 3 recalls, however, that difficulties began earlier. The Memorandum of Intent on Transboundary Air Pollution signed by the parties in 1980 committed each country not only to develop a bilateral agreement, but also to take interim steps to reduce transboundary air pollution. The Carter administration ignored this commitment when it sought relaxation of emissions limitations for coal-burning power plants in the Midwest. The move failed because of congressional opposition, but the negotiations with Canada suffered. At that time Canada's air pollution standards were much less stringent than those in the United States. However, since then Canada has unilaterally taken the steps to reduce emissions that it had hoped would be taken jointly. With the endorsement of the special envoys' report, the Reagan administration appeared to be committed to increasing its research efforts and taking some preliminary steps to reduce emissions. The Canadians have been repeatedly disappointed by the administration's actions since that time and see little prospect for an agreement until the U.S. position changes substantially.

In chapter 4 the authors retrace the activities of the bilateral research groups that were established under the 1978 agreement between Canada and the United States to prepare a common scientific basis for negotiations. The work was interrupted and politicized following the 1980 elections in the United States. Although most of the work was completed, the United States reneged on its commitment to complete publication of the final version. A lesson to be learned from this is that it may not be wise to assemble research groups composed primarily of government scientists, rather than to rely on the services of established science-policy institutions.

Chapter One
Acid Rain Is Different

The Questions We Asked

In this chapter we outline the issues that will be explored in depth in subsequent chapters. The following questions form the basis for discussion: What kind of policy is needed to deal with an environmental issue that has unique characteristics, is as yet imperfectly understood, and affects both Canada and the United States? Since there is disagreement between the two countries on how to respond to the acid rain threat, is there also disagreement in the assessment of the nature of the acid rain phenomenon and its effects?

This leads to our central question: How can neighboring countries with highly integrated economies, compatible political systems, and a history of peaceful resolution of conflicts go about influencing policy developments on the other side of the border when vital policy interests of its own are affected? Or conversely: What actions should be avoided by either side because they will be misunderstood and detrimental to a mutually acceptable outcome? So far, the Canadian government and people have shown more concern about the dangers of acid rain than the United States. As a result, Canada feels compelled to accelerate both the development of a domestic U.S. acid rain policy and the adoption of a bilateral agreement. What roles are available to Canada in influencing the development of U.S. policy? More generally, what is appropriate and helpful, and not counterproductive, on the part of either side in influencing a friendly neighbor's policy?

Our work leads us to the conclusion that acid rain is such a unique

---

This chapter was written by Jurgen Schmandt, Hilliard Roderick, and Andrew Morriss.

and complex environmental issue that it cannot be resolved satisfactorily within the existing policy framework for controlling air pollution. Moreover, currently contemplated extensions of existing policies by either Canadian or American proposals will not eliminate the problem. A broader policy response is needed. To support this point it is necessary first to consider what makes acid deposition an unusually difficult policy problem and then to identify the key elements of a comprehensive acid rain policy. Later chapters will provide details on specific policy options, in particular those that might be taken, jointly or individually, by various actors in the two countries to supplement formal negotiations between the two national governments.

The Difficulty of Policy Formulation

The central issue in the acid rain controversy depends upon the notion that sulfur dioxide (SO_2) and oxides of nitrogen (NO_x) released from factory smokestacks and motorized vehicle exhausts are primarily responsible for the creation of acid deposition.[1] It would seem reasonable, therefore, to expect that stricter controls of these pollutants would take care of the acid rain problem.

This assumption underlies the policy debate on acid rain, which is focused on control of SO_2 and, to a lesser extent, NO_x. The approach is valid up to a point. If new controls removed most of the two pollutants from the atmosphere, the problem of acid rain would most likely be resolved, except for the possibility that some irreversible damage may have occurred. But current proposals aim at reducing sulfur deposition only by one-half. Will this be enough, and will a cut in emissions achieve a similar reduction in deposition? Until recently, the linkage between emissions and deposition was unknown. The National Academy of Sciences has reported that this uncertainty has been removed. According to a report published in 1983, we can now conclude unequivocally that in eastern North America average annual emission of sulfur dioxide from power plants and other industrial facilities is roughly proportional to deposition of sulfate.[2] The conclusion removes a major uncertainty and reaffirms the central role of sulfur dioxide in the formation of acid deposition. Nitrogen added to watersheds, on the other hand, may act as a "fertilizer," with much of its acidity being removed in the process.[3] However, "acid shock" associated with spring snowmelt seems to be more serious in the case of NO_x. Other atmospheric pollutants that play a significant, but not well-understood, role include ozone and hydrocarbons.[4] Given our current state of knowledge, controlling SO_2 and NO_x emissions would appear to be an important first step in reducing acid deposition. More extensive measures may be necessary later.

Multiple effects on ecosystems have been observed. These effects can not be readily predicted. For instance, the buffering capacity of soils will often determine the pH of aquatic systems. In addition to chemical transformations that take place in the atmosphere, sulfates and nitrates interact with other chemical constituents after acid deposition has occurred. As yet not fully understood threats to human health may be implicated, as, for instance, from the increased solubility of metal ions at lower pH levels. Finally, the phenomenon operates on a regional, perhaps continental, scale. At the present time large parts of Canada and the United States are affected, and sources of pollution and damaged areas are found on both sides of the border. These factors make acid rain different from earlier forms of pollution. It is unlikely, therefore, that simple extensions of existing environmental statutes will resolve the acid rain problem in North America.

The new levels of complexity, as well as the enlarged geographical range, place acid rain in a new group of pollutants that will require new policies and institutional capabilities. Other examples of global pollution include the buildup of carbon dioxide in the atmosphere, the accumulation of pollutants in the oceans, and the depletion of ozone in the atmosphere. Continents, possibly the entire globe, are affected, making it necessary to develop control strategies and institutions capable of operating effectively in large areas that transcend political boundaries. At the present time no country or international organization is equipped—technically, organizationally, or politically—to address the new forms of pollution on the scale at which they operate. Controlling pollution on a continental or even global scale will be difficult for several reasons, including the technical complexity of the issues, the high degree of scientific and economic uncertainty, the lack of political resolve and authority, and the absence of institutional capability. Resolving the acid rain issue will be important for the next stage in environmental policy. Our policy response will set a pattern for other new forms of pollution that affect continents and the entire planet.

Contrasting Methods of Pollution Control

To illustrate the novelty and complexity of acid deposition, we shall contrast it with earlier forms of environmental pollution. In so doing we shall develop our argument that old forms of pollution control will play a role in resolving the acid rain issue but must be supplemented with new policy concepts and institutional capabilities.

Controlling Conventional Pollutants   Environmental policy began with programs to control "media-specific" pollutants, such as particulate matter and sulfur dioxide in the air, or bacteria and phosphorus in the water. The environmental statutes of the late sixties and early seventies designed to control these pollutants were based on three concepts: pollutants could be controlled *one substance at a time;* controls could be developed independently for *each environmental medium;* and that remedial action was needed primarily to alleviate *local* pollution problems.[5] This reflected an understanding of pollutants as fairly stable substances that occur in a single medium and affect small areas—cities and metropolitan areas, and in extreme cases small states. Air pollution, which previously had been viewed as a nuisance, was now recognized as a risk to human health that could aggravate existing health conditions, particularly for the at-risk groups —children and the elderly. Using this conceptual model of pollution, the federal government controlled automobile exhausts, established ambient air standards for several pollutants (among them SO_2 and NO_x), and set emission limits for new stationary sources. Policy was initiated at the national level, but the states became responsible, under federal supervision, for implementing most of the new control programs.

Current attempts to control acid deposition by reducing SO_2 emissions rely entirely on the legal framework created for the control of conventional pollutants. In our judgment this provides too narrow a base for dealing with acid rain. The Clean Air Acts in the United States and Canada were designed to control single pollutants in local environments. This local orientation ignores long-range effects. It can even make them worse, as in the case of the "tall-stacks" policy (no longer permissible in the United States), which was used by industry to meet emission standards close to the ground. Proposed acid rain amendments to the Clean Air Act expand the area of control, at an estimated cost of several billion dollars annually. The controls will reduce the severity of acid rain and the size of affected areas, and will, in addition, reduce conventional air pollution. But acid rain is unlikely to disappear even if these measures are adopted and implemented.

Controlling Toxic Pollutants   A second environmental policy model emerged during the seventies when it was recognized that many industrial products and processes release toxic substances into the envi-

Acid Rain is Different 11

ronment. Almost all substances can prove toxic to humans if ingested or inhaled in sufficiently large quantities. Health and environmental policies are concerned with those toxic substances that cause long-term and irreversible disease (such as cancer), birth defects, and genetic or behavioral damage. Detection and control of this class of pollutants are complicated by many factors, and their effects may take a lifetime to discover. Scientific evidence about the nature and health effects of pollutants is difficult and expensive to produce. Vast amounts of scientific, technical, and economic information must be generated and interpreted. While much has been accomplished since the early seventies, information about toxins in the environment remains incomplete. The number of possibly dangerous substances is large—some 65,000 chemicals are in commercial use,[6] of which about 5,000 have been listed as having some inherent hazard potential, but detailed estimates of risk have been prepared only for a few hundred substances. Toxic substances migrate from one medium to another and tend to have synergistic and additive effects. Direct links between cause and effect are difficult to establish, and statistical probabilities and models must be relied on to guide policy. Moreover, the concept of safe standards is no longer applicable; decisions need to be based on risk estimates instead. Evidence of this kind is more open to disagreement and controversy than direct cause-and-effect relationships.

To supplement the media-focused model of environmental regulation with programs designed to control toxic pollutants, Congress took action along several lines. The Toxic Substances Control Act, aimed at control of industrial chemicals, was passed in 1976. Toxic substances legislation for pesticides, drinking water, and industrial wastes was passed or updated. Special toxic substances provisions were incorporated in the air and water statutes. Control of toxic substances in the environment became primarily a responsibility of the federal government, assigning only a limited and less well-defined role to the states.

There is evidence that acid deposition frees toxic metallic substances in the soil, creating the potential for contamination of drinking water and edible fish. Therefore, toxic as well as conventional pollutants are implicated in the acid rain phenomenon. Yet proposed acid rain controls do not address the toxicity issue, nor do they deal with other possible health effects. A comprehensive acid rain policy will need to integrate aspects of both policy models.

Scientific and Economic Uncertainty

How can policy be formulated when many questions regarding acid rain remain unanswered? A few examples will illustrate the difficulties standing in the way of informed decision making. For instance, multiple sources of pollutants and transformation of pollutants during transport make it difficult to establish firm links between causes and effects. It is also possible that the danger from acid deposition is more far-reaching than the current policy debate suggests. A 1983 review of the state of knowledge on acid rain by a panel of scientists for the White House science-policy office raised questions about the effect of acid rain on microorganisms in the soil. The report suggests that increased acidity is perturbing populations of denitrifying microbes, upon which the entire biosphere depends. Even though it may take many years to establish such an effect, the prospect of such an occurrence is grave.[7]

Most experts agree that this particular threat from acid deposition is not yet supported by strong evidence. But the case illustrates that current knowledge is insufficient to grasp the full ramifications of acid rain and its effects. An appropriate policy response must recognize this fact and allow for experimentation and evolutionary change.

Uncertainty has played a major role in the acid rain policy debate. Uncertainty over the causes is cited as a reason for waiting to institute controls until more is known. Uncertainty over the efficacy of various control methods is used to justify research programs to develop new control technologies. And uncertainty over the potential severity of effects is used to argue both for and against immediate control measures.

It is important to recognize the variation in types of uncertainty, and to distinguish between them. Figure 1-1 shows the relationship between scientific uncertainty and political uncertainty. Scientific uncertainty exists where, given a set of data, scientists differ on the correct interpretation of the information. Political uncertainty arises from trans-scientific disputes. Where the two overlap, the issue becomes confused. Scientific evidence can be used inappropriately, and political values can color data interpretation. This happened, for example, in the dispute over the bilateral Memorandum of Intent on Transboundary Air Pollution (MOI) work group reports.[8] Scientists from the two countries agreed on the summary of scientific evidence. But the U.S. scientists associated with the assessment felt that the conclusions presented by the Canadians overstepped the available data. The Canadians, for their part, saw American caution in drawing conclusions as a reflection of political values.

Figure 1-1.   Scientific and Political Uncertainty

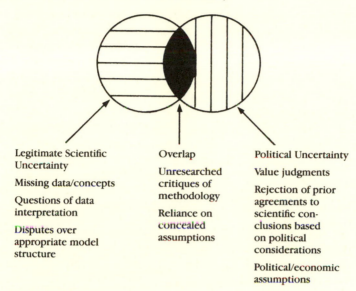

Legitimate Scientific
Uncertainty

Missing data/concepts

Questions of data
interpretation

Disputes over
appropriate model
structure

Overlap

Unresearched
critiques of
methodology

Reliance on
concealed
assumptions

Political Uncertainty

Value judgments

Rejection of prior
agreements to
scientific con-
clusions based
on political
considerations

Political/economic
assumptions

Uncertainty is involved in four types of interactions (see figure 1-2). There is uncertainty in interactions among scientists because of differing interpretations of assumptions, theories, and data. There is uncertainty between scientists and policymakers because policymakers often fail to grasp the extent and the limitations of scientific knowledge. For example, many policymakers assume that because scientists cannot identify specific sources as causing specific increases in acid deposition, they cannot also make relevant statements about general principles of acid deposition.

The flow of uncertainty between scientists and policymakers moves in both directions. Scientists often have difficulty understanding the political uncertainties that are imposed. The disclaimers added to the MOI work group reports by the U.S. delegation made it difficult for the scientists involved to understand their role. If political considerations can overrule their work, then how can they contribute to the policy debate? Policymakers, for their part, suffer from uncertainty of a different kind: Do they understand the problem? Can they assess the significance of scientific results? What values are determining their choice of facts in searching for solutions? How will their actions be viewed by political allies and enemies, or by voters and supporters? An example is the inability of some of the U.S. and Canadian policymakers to agree on basic

Figure 1-2.   Interactions Producing Uncertainty

Scientist   ⟶   Scientist
                    Incorrect transmittal of data, methodology,
                    or assumption

Scientist   ⟶   Policymaker
                    Inability to understand analysis
                    Misperception of conclusions

Scientist   ⟵   Policymaker
                    Rejection of conclusions due to
                    political considerations

Policymaker   ⟶   Policymaker
                    Failure of communications

facts, such as increased emissions due to increased economic activity.

Some of these uncertainties stem from the inherent subjectivity of scientific analysis. Economic interests tend to emphasize different aspects of uncertainty and disregard areas of knowledge seen as unfavorable to their position. Other uncertainties result from disagreement on how to proceed in the face of the other unknowns—how should we deal with the uncertainty of the effectiveness of control technologies? Should we proceed now or wait for possibly more effective controls to be developed? Value decisions and political uncertainty often enter into the resolution of these questions, sometimes masquerading as scientific uncertainty.

We suggest that some uncertainties should be dealt with by having the policy response take them into account. Instead of passing an acid rain control statute that assumes full knowledge about the problem and its solution, legislative action might better take the form of authorizing administrative controls on a temporary basis, with the requirement of congressional or parliamentary evaluation and reauthorization review after a designated period of time. The imposition of control requirements is another good example of dealing with uncertainty in the policy response. If SO_2 sources are required to install scrubbers to control their emissions, then an additional layer of uncertainty is added, since the effectiveness of the mandated controls would be unknown. However, if a percentage reduction in emissions is required, rather than a specific technology, then some of these uncertainties are avoided, since industry remains free to adopt new technologies as they become available.

Uncertainty is not new to environmental problems. Why, then, does it pose new difficulties with acid rain? First, uncertainty exists at many levels. Second, the amount of uncertainty imposed by political and economic concerns, and in the scientist-policymaker relationship, is greater. Third, uncertainty is used to argue against the imposition of controls generally and not just to dispute specific regulations or requirements.

At the same time, the nature and extent of uncertainty concerning acid rain should be judged against the background of what is already known. Without scientific research into the causes and effects of acid rain, the problem would never have been identified and defined in terms that make rational debate and the search for policy possible, and a common ground for the political debate would be missing entirely.

The appropriate role for scientific evidence in regulation is not to provide answers concerning the preferred policy response, but to give the decision maker a meaningful problem definition, to formulate and test plausible hypotheses concerning cause-and-effect relationships, and to make the scientific findings available in terms that can be understood by nonexperts.[9] Judging from this perspective, we know enough to recognize acid rain as a serious environmental problem that is associated with industrial activities. The claim that any kind of policy response would be premature ignores the amount and quality of available knowledge and fails to understand the role of scientific evidence in decision making. The opposite claim, that final answers are in hand now and permanent solutions can be instituted, would be equally wrong. The nature and extent of our understanding of the issue—what we know as well as what we do not know—support our call for a step-by-step policy response to acid rain.

The uncertainties involved in acid rain have been linked to the most basic issues; namely, whether SO_2 and NO_x emissions cause acid precipitation. But this question can no longer be presented as unresolved in the policy debate. We found solid agreement among experts from both countries that burning of fossil fuels was at the root of the acid rain problem in North America.[10] But fundamental disagreements remain. Unlike previous instances of air and water pollution, links between specific sources of emissions and damages are difficult to establish. Thus, the uncertainties are least with regard to definition and largest in respect to resolution of the acid rain issue.

Under such circumstances the question of cost becomes critical —cost of inaction as well as of alternative controls. The dispute over acid rain is as much a question of allocation of costs as it is of technical solutions. Most earlier environmental issues did not address the question of costs at their outset, but began with a strong policy commitment to clean up the environment. Several environmental acts passed in the early seventies went so far as to forbid the regulatory agency to take cost into consideration. With acid rain the issue of cost arose at the outset. Unlike the debate over the scientific foundations of acid rain, the economic questions—How much damage is caused? Who bears the costs? What will controls cost? Who will pay for controls?—have remained largely unanswered. Available cost estimates vary widely. Even for the limited question of how much it would cost to reduce SO_2 emissions in the United States by ten million metric tons, cost estimates vary from $2.7 to $5.3 billion (in 1985 dollars).[11] U.S. legislative proposals to control acid deposition would cost about $3–$6 billion per year.[12] And costs are not only measured in dollars, but in jobs and votes as well.

The lack of consensus on the costs of damage and controls reflects a lack of consensus on the questions that need to be asked and the assumptions that should underlie the analysis. Without agreement on common assumptions, it is impossible to conduct a meaningful evaluation of the costs of acid rain damage and control. In a study prepared for the U. S. Senate Committee on the Environment and Public Works, Larry Parker suggests that a minimal degree of analytical comparability would be reached if the following cost factors were considered consistently: (1) the amount of SO_2 and NO_x removed; (2) the method of allocation of the reductions required; (3) the means used to obtain reductions in emissions; and (4) the nature of the affected industries.[13] In the absence of a consensus on questions and assumptions, uncertainty about economic implications of acid rain is probably a more serious impediment to political action than scientific uncertainty.

The great uncertainties involved suggest several implications for policy development. First, despite the increased knowledge more research will bring, it will not resolve the political controversy, and it should never be expected to do so. At best, it will reduce the scientific uncertainties and clarify their overlap with political questions. Thus, while more research will be necessary for informed decision making, it will not alter the need for difficult choices among political alternatives. Second, the political nature of the decisions that need to be made must be clearly identified both to the decision makers and to the public. Third, expansion of the dialogue between policymakers of both countries can promote development of policies to resolve the problem of acid rain.

The call for more research is almost universal. It does not, however, offer the possibility of discovering "the answer" to the policy questions of acid rain. Even if, and this is most unlikely, all the "pure" scientific uncertainties were resolved tomorrow, the economic and political questions would still remain. Although research can help guide policymakers part of the way toward resolving these uncertainties, it is no substitute for hard policy choices.

The decisions to be faced involve choosing between interest groups —between the Northeast and the Midwest, between the coal interests and the forestry interests, between electrical power and leisure industries, between short-term economic gain and long-term preservation of our natural resources. These choices are political ones and resolving them will require unusual doses of determination and sagacity.

Policy Options

If, as we conclude, old policies and concepts of pollution control cannot by themselves resolve the acid rain issue, it follows that most of the hard work involved in developing a comprehensive control strategy remains to be done. At the time this book was first written, neither the United States nor Canada had in place statutes or programs specifically designed to deal with acid rain. Canada had proposed that both countries reduce SO_2 emissions by 50 percent as part of a bilateral agreement to reduce acid deposition. Although this is not the kind of comprehensive program that we foresee will be necessary to address the problem adequately, it would have been a first step in the right direction. However, in the absence of a positive response from the United States, Canada in 1985 unilaterally implemented a program to reduce SO_2 emission from stationary sources by 50 percent and bring automobile emission standards into compliance with those in the United States. These provisions will address about 50 percent of the acid deposition affecting eastern Canada. Now it is time for the United States to act. Until acid rain legislation is passed, environmental control agencies have no mandate to curb acid deposition. This point is fundamental. We do not yet have workable policies to deal with the type of pollution represented by acid rain. Large areas are affected; pollutants are transported over hundreds of miles; continental, perhaps global effects are created; environmental as well as health concerns must be considered; and both media and toxic pollutants (e.g., in the form of toxic metals leached into surface waters) are involved, but their interaction is not well understood. Other gaps in knowledge exist with regard to the chemical transformation of pollutants during transport. These include effects on crops and forests, possible threats to human health, and the exact relationship between sources of emissions and receptor areas.

Current Plans

Plans to deal with acid rain, both in the United States and Canada, are extensions of existing air pollution controls, and will work primarily by means of further reducing permissible emission levels for sulfur and nitrogen oxides. Amendments to the Clean Air Act and several acid rain bills considered by the U.S. Congress in the early 1980s would have introduced SO_2 controls for the United States east of the Mississippi.[14] Adoption of regional environmental standards would be a significant departure from uniform national standards, which have been the mainstay of environmental policy over the last two decades.[15] The economic, social, and political implications of such a shift would be serious. It has become apparent that the establishment of what is perceived as regional inequities is politically infeasible. More recent amendments provide for nationwide controls and equalizing measures, such as a fund to subsidize the capital cost of controls in order to cap residential electric rate increases. Regional taxes and subsidies, however well intended, would create new economic and social distortions.

In trying to defend its position of no action, the United States has often used the argument that Canada's environmental controls are less stringent than those in the U.S. For instance, when Canada proposed a 50 percent across-the-board cut in SO_2 emissions in both countries, it was claimed that Canada's emission levels would still exceed those in the United States. In addition, until recently, automobile emission standards in Canada were much less stringent than U.S. standards. It is apparent that Canada has taken this criticism to heart. Recent federal action will bring automobile emission standards into compliance with those in the United States, and the seven eastern provinces have each enacted legislation that will result in an overall reduction in SO_2 emissions from stationary sources by 50 percent. This move was intended not only to reduce acidification due to local emissions, but to show the United States that Canada is serious about addressing the overall problem. However, because nuclear power in Canada is a realistic alternative to coal-based electricity, utilities in the United States claim that it is much less costly for Canada to implement such policies. What will happen if Canada follows the United States, and public opposition to nuclear power severely restricts growth of this industry? Conversely, might the threat of acid rain damage lead to a reconsideration of the nuclear energy option in the United States in the 1990s?

The European approach is also focused on controlling emissions of sulfur and nitrogen oxides. Guidelines for reduced emission levels have been issued by international organizations, but no comprehensive international control program exists, nor is one likely to be adopted. Any legally binding control measures will depend on action at the national level. Responding to reports of damage to forests, the West German government, for example, has decreed lower sulfur dioxide emissions from stationary sources. The installation of catalytic converters to control NO_x emission from cars is not yet mandatory. But even these limited measures represent a considerable change in policy. For years Germany (together with the United Kingdom) had resisted Scandinavian initiatives against acid rain. Since existing as well as proposed European emission standards for SO_2 are more permissive than U.S. standards, little guidance for policy development in North America can be expected from Europe. The same is true with regard to reduction of NO_x emissions.

Domestic versus Bilateral Solutions

The current initiatives in the United States and Canada have in common that they focus on one or two major pollutants (SO_2 and NO_x) and attempt to control acid rain under existing air pollution statutes, which were designed to control local air pollution. The proposed controls do not consider the fact that much of the danger of acid rain (for example, the damage to soils or drinking water) may result from the interaction of SO_2 and NO_x with toxic pollutants and from complex chemical processes that occur during the long-range transport of the pollutants. Both governments, in our view, need to broaden their view of acid rain and recognize the issue for what it is: a problem of unprecedented complexity with many aspects that are not yet understood, with little precedent to guide action, and with powerful economic interests that see their livelihoods threatened in opposition. If that much is agreed upon, it becomes clear that both countries need more than an expanded version of the current Clean Air Acts in Canada and the United States. In searching for a comprehensive acid rain control strategy, what should be the appropriate division of labor between national and bilateral efforts?

In the process of developing policy, increasing the dialogue between policymakers and representatives of different interests would help. Canada and the United States share their environment. National policy in each country affects the other. Informal dialogue offers the opportunity for both nations to make their ideas and concerns known without the constraints of formal negotiations. But whatever will be achieved between the nations will have to be based on policy choices made at home. Without sound acid rain policy at the domestic level, little can be accomplished internationally.

Past experience in addressing environmental disputes between Canada and the United States suggests that bilateral actions will play a useful but limited role; they are likely to supplement domestic initiatives but unlikely to become the driving force for resolving the acid rain issue. We make the assumption, therefore, that no full-fledged international control policy will emerge, and that domestic policy initiatives will have to lead the way. But within the framework of enlarged national policies, cooperation between the two countries (and, eventually, Mexico) must be agreed upon and implemented to an extent that far exceeds current political will, experience, and institutional capabilities. Specific measures include joint research, monitoring, control experiments affecting large areas, and harmonization of national policies.

Domestic decisions will do more than decide between control of acid rain and continuation of the status quo. It is also likely, given the differences between the two countries in size of population and gross national product, that decisions by the United States will determine the outcome of the issue. We expect that decisive domestic action will be delayed until the perceived damage is serious enough to generate broad support for another "grand leap" in environmental policy. While such support seems to exist in Canada, the same is not yet the case in the United States. The fear of serious damage observed elsewhere has been the prime motivation for protection of environment and public health in the past, and this pattern is likely to continue in the case of acid rain. Once people become genuinely concerned about the effects of acid rain on wildlife, vegetation, and human health, political momentum will build up quickly.

In the meantime, Canada is limited in the number and kind of measures it can take to facilitate U.S. policy developments, especially given the current economic and political climate in the United States. Some measures carry considerable risk, such as support of particular legislative initiatives in the Congress or the building of grass-roots constituencies. Congressional as well as interest group coalitions tend to change frequently. Too close an identification with particular measures may reduce the weight of the Canadian argument in formal negotiations.

Strategies for Domestic Policy Development

Three strategic options for the United States and Canada are discussed in the current policy debate: (1) do nothing; (2) conduct research; or (3) conduct research and impose emission controls. What are the implications of each of these options?

The do-nothing approach has been justified with the assertion that natural phenomena, not industrial activity and urbanization, caused acid deposition in Europe and North America. For all practical purposes this position is no longer tenable in the light of overwhelming scientific evidence to the contrary. While the do-nothing approach found support in the early days of the first Reagan administration, responsible government officials have at last recognized the seriousness of the issue and the need for a much-increased research and monitoring effort.

Conducting research, however, is not a single action. Is research to be conducted by the governments, by contractors, or through grants to university faculty? Will the source of funds be government, the private sector, or both? How can industry be convinced that its own funds will be well spent and serve its own best interests? The answers to these questions may determine the outcome in some instances, and will affect, at the least, the speed with which various answers are reached.

Similarly, emissions controls can take a variety of forms, from a simple technology-based standard (e.g., a requirement that all sources install scrubbers), to a media quality approach (e.g., uniform reduction of emissions by fifty percent). The Congressional Office of Technology Assessment has estimated that permitting electric utilities to devise their own methods to control sulfur dioxide and obtain the same levels in reducing emissions could save about $1 billion a year over mandated installation of scrubbers.[16] While interest groups may agree on general issues surrounding acid rain, they may differ on strategy and timing. For example, the agreement between Canadian industries and environmental groups that acid rain should be controlled does not necessarily extend to these more specific issues. Such differences could be exploited by the opponents of a particular measure to prevent the development of a coalition behind it. Or they could be used to force an unholy alliance between those in favor of maximum measures and others opposing all controls.

It is unlikely that the current air pollution controls and research efforts will be reduced. Consequently, the spectrum of policy choices is constrained so that maintenance of the status quo is the least we

Figure 1-3.   Spectrum of Policy Choices

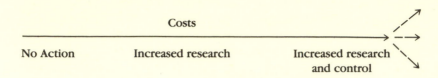

Costs

No Action   Increased research   Increased research
                                               and control

can expect (see figure 1-3). At the other end of the spectrum, additional controls are not a uniform option; they encompass a wide variety of measures, all with differing implications. Thus a more accurate portrayal of the policy options would be (1) increased research with no additional controls; (2) additional research with moderate expansion of controls; or (3) increased research and ambitious new controls (see figure 1-4).

Several factors must be considered when choosing among policy alternatives, including reversibility of effects; prospects for development of alternative control measures that might be either more effective, or less expensive, than current control measures; and the question of who should pay for controls.

Within this spectrum of policy choices, the Canadian role might be to awaken sympathetic U.S. interests to their own stake in acid rain. This role stems from Canada's view of the problem as now largely in the political arena, with most of the scientific questions necessary for action sufficiently settled for controls to proceed. Canadian action in the United States is required, in this view, by the unwillingness of the United States to respond to Canadian concerns. It would be important, in particular, to involve economic interests such as the forestry industry in the debate. Also, areas of the United States outside of the Northeast should be helped to recognize their own interests in curtailing acid rain.

The Canadian government supported this activist policy until the change in administration in 1984. This policy was criticized by some American industry and government representatives—among them the leaders of the Reagan administration—as exercising what they considered an inordinately strong influence on the U.S. policy-making process (through lobbying, appearances before Congress, and similar activities). This, they charged, represented interference in American domestic affairs. The critics suggested that Canada, in order not to overstep the boundaries of diplomatic propriety, should act as a "more gracious loser" in its efforts to influence U.S. policy. This might include

Figure 1-4.   Modified Spectrum of Policy Choices

Costs

——→

| Additional research/
Existing controls | Additional research/
Moderate expansion
of controls | Additional research/
Extensive expansion
of controls |

allowing U.S. citizens greater access to Canadian courts for environmental suits, or reducing Canadian emissions before reaching an agreement with the United States.

Proponents of Canadian "interventionism" counter with the assertion that such actions are irrelevant to their goal of achieving a meaningful reduction in U.S. emissions of acid rain precursors, which they feel is the primary source of acid rain in Canada. They reject the notion that by being a shining example, Canada could exert moral pressure on the United States that would be sufficient to lead to significant U.S. actions. (This appears to be the case; the enactment of emission control measures in Canada has not elicited a similar response in the United States.) Canadians also argue that the effectiveness of Canadian lobbying on acid rain legislation in the United States had been greatly overstated by their critics. They point out that their presence, and the overall presence of procontrol groups, was far less than the presence of utility and other anticontrol groups. Moreover, they cannot provide legislators with either of the traditional inducements for support—campaign support and votes.

The U.S. view, as expressed by the Reagan administration, reflects a general feeling that the role of a foreign government, even one that is a friendly neighbor, does not include intervention in what the United States regards as a domestic policy dispute. Thus, the issue of whether this is an essentially domestic policy dispute or primarily a bilateral issue is important in determining the appropriate level of Canadian involvement.

The Range of Options Is Narrowing   Several important conclusions follow from the above discussion. First, research on acid rain must be structured to address those issues that are crucial to resolving the policy debate. The identification of the "critical path" to the resolution of those issues should be a high priority for any research effort. Second, it is inappropriate to view emissions limitations proposals and cost allocation schemes as alternatives, because they address different issues. Third, there is increasing agreement by the policy community on the range of policy options available to the two countries. There remains disagreement, of course, on how to resolve the issues. Nevertheless, the wait-and-see approach has been dismissed by all responsible parties, and the agenda for the acid rain debate is now largely shaped by the proponents of greater control rather than by those advocating lesser controls. Given the current political and diplomatic climate, existing controls and research levels are unlikely to be reduced. The controversy surrounding the Reagan EPA, and the Canadian determination to raise the issue, have collaborated to constrain the spectrum of policy choices (see figures 1-3 and 1-4). The best the proponents of less control can achieve is a postponement of more rigorous controls, and not the extensive relaxation of current controls that they had hoped for after the 1980 elections.

We found that the narrower range of options, as illustrated in figure 1-4, is supported by Americans and Canadians alike. This represents an important step forward in the resolution of the debate between the United States and Canada. Debate over specific policy proposals within that range can be undertaken. This developing consensus is the result of the dialogue within the policy community over the issues in the acid rain debate since 1978.

Finally, perhaps the most important conclusion about policy options is that domestic policy development, in the United States as well as in Canada, must carry the principal load for finding workable solutions to the acid rain problem. This does not mean that international actions will not be important, only that the progression will be from national to international initiatives. It is thus mandatory that the domestic policy responses of the two countries, while developed independently, are coordinated with each other. Otherwise the international character of the acid rain phenomenon risks becoming a continuing source of friction between the two countries. The cost of not resolving the dispute could be high, not just in terms of continued damage from acid rain, but also because relations between the two countries would deteriorate generally and the resolution of other issues would become more difficult. While both sides have been careful not to link the acid rain question to other aspects of U.S.-Canadian relations, such linkage might become automatic over time. On the other hand, progress in resolving the acid rain dispute will improve relations between the two neighbors. It would also entail bilateral action on a scale unprecedented in environmental policy.

Joint control of pollution in the Great Lakes involved action covering a large watershed. Yet, the area is small compared to the continental scale of the acid rain problem, and no new national legislation was required to deal with pollution of the Great Lakes. Domestic law in both countries was used to support bilateral activities. An existing international body, the International Joint Commission, was used to take on new investigatory and monitoring responsibilities.[17] The institutional model of the Great Lakes commission was later used in Europe to control pollution of the Rhine, so far with limited success. The Commission of the European Economic Community has recommended reduction of SO_2 emissions to its members, but these measures have not been translated into effective national or international policies.

We do not know of large-scale environmental programs anywhere in the world that involve integrated control activities on the territories of two or more nations. An effective North American acid rain program will have to break new ground in finding workable institutional solutions. Whether this process should be built around an expanded International Joint Commission or should involve new institutions is beyond the scope of this study. However, both countries would be well advised to recognize the need for joint institutions for conducting research, monitoring pollution, and testing control strategies.

Yet the bilateral policy response will continue to be driven primarily by domestic forces. The Canadians acknowledge this in their hope to affect U.S. policy by involving U.S. interests in the debate. The United States acknowledges this through its emphasis on the Clean Air Act reauthorization as the primary vehicle for addressing the problem of acid rain, rather than emphasizing a bilateral agreement. If they are to succeed, however, these domestic policies must be built around common policy objectives and programs, and not merely around common technical tasks. Components of an effective policy must include a fully integrated monitoring system, joint demonstration projects, policy-oriented hearings and conferences, policy and economic analyses, a bilateral coordinating commission, and a constant dialogue between policymakers, between scientists, and between both groups. Considering options like these would open the door for a broadly conceived acid rain policy in North America.

Components of a Comprehensive Policy

A comprehensive acid rain policy, in our judgment, will not become politically acceptable until extensive damage, and fear of further damage, mobilize broad public support. The Reagan administration has opted for a strategy that delays new controls while research into the different aspects of the acid rain phenomenon proceeds at an accelerated pace. Our work supports the need for a strong acid rain research program. We also see the need for additional reductions of SO_2 and NO_x emissions. But even if this were done now, some of the potentially most harmful effects of acid rain would not be addressed. A comprehensive and long-range policy, therefore, needs to entail three components: (1) a strong research and monitoring program; (2) additional controls of SO_2 and NO_x; and (3) a firm commitment of the resources needed to develop and test longer-term solutions. Work on all three aspects of the policy should proceed in parallel.

Planning and implementation of such a policy should entail a much wider range of actors than the two federal governments alone. States and provinces, professional and scientific organizations, labor unions and industrial associations, courts, and environmental and other public interest groups should also be involved. But decisions and agreements, obviously, require government action. Here the key question is to what extent the federal governments of the United States and Canada can formulate sound policies to control acid rain under conditions of uncertainty. Damage to the environment has been established but many effects are not yet well understood. The driving force behind the need for policies to control acid rain is the public perception of damage already done to the fish and lakes of New England, Ontario, and Quebec, and the fear that continued exposure of the northeastern environment to acid rain may produce damage to crops and forests that will significantly reduce their economic value, as well as the possibility that human health may ultimately be affected. The potential damage cannot be quantified at the present time because too little is known about the effects of acid rain over long periods of time (decades or even centuries). The costs of acid rain control cannot be estimated with certainty because all of the factors involved in producing the phenomenon are not known with reasonable certainty, and practical means of control may need to be reconsidered as knowledge grows.

How, then, can policy be formulated to deal with a phenomenon that affects both Canada and the United States, that so far has caused extensive but limited environmental and economic damage to both countries, and where the presently known means to limit future material damage are estimated to be costly to the economy of a region of the United States (the Ohio valley) that is not expected to benefit directly from control of acid rain?

It has been estimated that scrubber control of coal-fired power stations that would reduce SO_2 emissions by 50 percent will cost $3– $6 billion a year in the United States. In the first edition of this book we advocated spending perhaps 5 percent of that amount each year to examine the nature of the problem and test alternative control strategies. This would amount to some $150–$300 million a year for the next five to ten years, approximately five times more than the funding levels committed at that time. (The 1987–88 budget for the National Acid Precipitation Assessment Program is $85 million.) We felt that this would allow for increased research and monitoring, and broaden the scope to include testing of regulatory options. It would show the public and industry, as well as our neighbor, that the problem is not being neglected, but also that no hasty decisions, in addition to the controls already required under existing law, are being made by the U.S. government. Since that time bilateral negotiations between Canada and the United States have led to similar conclusions.[18] However, that alone is not sufficient. Much has been learned in three years, and there is nothing to suggest that the problem is any less serious than previously thought.

Even though the Reagan administration is still reluctant to admit that control measures should be implemented now, it has advocated increased research on new technologies, in particular those designed to enable coal to be burned cleanly. In negotiations with Canada the administration agreed to initiate a five-year, $5 billion control technology commercial demonstration project, funded jointly by the government and industry. However, the money has not been appropriated by Congress and so far only $700 million has been allocated to the Department of Energy for such projects. Although such technologies could ultimately be the answer to the problem, both in terms of technical desirability and economy, this strategy could still be interpreted as a means of delaying the implementation of control policies. We feel that the development of a comprehensive policy is long overdue. The following considerations would guide development of such an acid rain policy:

1. The nature of the problem of acid rain requires a different policy and administrative approach than that used for local air pollution control. A different approach is needed because of the nature of acid rain production in the atmosphere, its transport over long distances (hundreds to thousands of miles), the variability of its direction of movement and deposition (due to seasonal variations in wind patterns and climate), and the difficulties of determining multiple initiating sources and relating their source strength to the strength of the acid precipitation.
2. Acid rain may lead to a slow, large, regional (ten million square

miles) environmental change of uncertain consequences. Recognition of the long time needed for effects to become significant —perhaps fifty years for forest products[19]—and their possible irreversibility means that the acid rain problem may be considered similar in some ways to the soil erosion and dust-bowl problems of the 1930s.

3. Effects on vegetation and human health, perhaps as a result of toxic substances in soil and drinking water, need to be better understood in order to integrate them into an appropriate control strategy.

In sum, policy concepts and mechanisms are needed to control pollution that is transported over long distances, affects large areas, causes multiple effects over long periods of time, involves different types of pollutants, and generates cross-media effects. The policy response to the acid rain problem should bear this in mind and *not* seek a quick administrative or technical fix. The potentially serious, long-term, and large-scale consequences of acid rain require that the governments of the United States and Canada face the acid rain problem as a growing threat to their economies and the environment of the North American continent.

Seen from that perspective, a bilateral agreement to control acid rain may initially have to center on common activities aimed at researching and monitoring acid deposition and testing alternative control strategies. At the same time both the United States and Canada need to take steps to bring domestic environmental policies in line with the new situation created by acid rain. This domestic response to acid rain should move in two directions.

First, as a short-term strategy, each government should ensure that existing controls for new stationary sources are complied with fully and without additional delays. Legislation may be needed to curtail techniques that have been used by industry to circumvent or delay the applicability of stringent standards for new stationary sources. This may apply, for example, to the practice of constructing ever-taller smokestacks or continuing operation of old sources far beyond their normal life span.[20]

A second step might involve imposition of the kind of controls now under consideration in Congress that are aimed at reducing sulfur dioxide emissions in the United States. Alternative measures, such as washing of coal, might also be considered—the technique is much cheaper and can achieve significant emissions reductions, but creates a new disposal problem for the residue from washing. Canada, for physical as well as political reasons, has the option to shift an increasing part of its electricity demand to water and nuclear sources. But it is also developing new potential sources of acid rain in the western provinces

The longer-term strategy must be based on a better understanding of the acid deposition problem in its different ramifications. This will call for a level of effort in research, monitoring, and testing that can be expected to be considerably larger and more multifaceted than the present assessment program. It will take a relatively large amount of money, professional effort, and time to find out about the nature of the problem, to monitor its development over time, and to define and test control strategies.

What we are proposing here is a large-scale planning, research, and demonstration effort by all interested parties to *consider alternative regulatory concepts and control strategies*. This is more than a traditional research and development task. We shall develop the point by considering the role of the U.S. Environmental Protection Agency (EPA) in controlling acid rain.

Up to now the EPA has dealt with acid rain only as a research and analysis task, not a current regulatory issue. This means that the country's central agency for environmental protection conducts acid rain studies but does not focus its resources and institutional capabilities on the implementation of an acid rain policy. EPA's large air office, as well as its program offices in charge of toxic substances and drinking water, have no mandate yet for dealing with acid rain. There was a period of several months, following the appointment of William Ruckelshaus as EPA administrator, during which the agency attempted to develop an acid rain policy for the Reagan administration. This effort, however, failed in the face of opposition from the Office of Management and Budget and other cabinet agencies.

Such a limited role on the part of the EPA will strike the reader as surprising, given the appearance of intense political activity accompanying the acid rain debate in recent years. But the restriction accurately reflects the fact that regulatory development to control acid rain has not been authorized by Congress. The rule-making process, which makes regulatory agencies responsible for moving from congressional policy directives to actual regulation, has not been set in motion to address the acid rain problem. It is only through rule making that the EPA can move beyond research and analysis and begin to develop and test control strategies. This points to a potential defect in our regulatory system: Regulatory agencies can begin the rule-making process only after Congress has told them, in general terms, what to do. This is a time-honored principle meant to safeguard the constitutional separation of powers between the policy-initiating legislature and the policy-implementing administrative agency. But the underlying assumption for such an arrangement is that Congress knows enough to develop policy and direct administrative action.

This arrangement contrasts with that in Canada, where the civil service has the primary responsibility for writing legislation. Each system has its merits. It can be argued that the EPA should be more involved in policy-making because field experience is needed to write a detailed acid rain statute; a more flexible temporary authorization by Congress would be more appropriate for new, complex, and ill-defined issues like acid rain. In this way policy could be developed in successive stages, with the initiative shifting between the legislature and the regulatory agency. It would replace a situation in which Congress attempts to write highly detailed legislation at the outset, and finds it difficult later to make adjustments in light of new information. Such an arrangement would have far-reaching consequences for the respective roles of Congress and the regulatory agencies in dealing with technically complex policy issues.

We realize that a temporary authorization would be difficult to implement and would give rise to many problems. But the question of how to tap the expertise of regulatory agencies in policy development deserves to be reconsidered for the new kind of environmental and health issues that confront us in increasing numbers. However, there is good reason to believe that, without more progressive policy directives, this strategy would not further the cause of emission controls and reduction of acid deposition. Because the EPA has not enforced the existing provisions of the Clean Air Act to the fullest extent possible, one of the objectives of those advocating stricter controls is to make each provision more explicit during the course of the reauthorization of the act. It is unfortunate that this administration's EPA is no longer perceived by many as a protector of the environment. The EPA has repeatedly testified before Congress that acid rain is not a serious problem and that existing statutes provide sufficient environmental protection to prevent harmful levels of acidification. Thus, although EPA does not have a formal role in the crafting of legislation, it is not without influence. In addition, there is often significant opportunity both to fine-tune the provisions of legislation during the rule-making process, and to more narrowly interpret its intent during implementation. At this juncture, congressional action on acid rain is the only way to make significant progress.

||||||||

||||||||

Chapter Two
The Nature and Effects of Acid Rain:
A Comparison of Assessments

||||||||

Introduction

Policy-making depends upon the availability of an adequate scientific data base. In the case of acid rain the subject is particularly complex, and policymakers must rely heavily on scientists to analyze its nature and effects. It is important to ask, therefore, whether policy formation has been delayed because of insufficient or conflicting evidence. Major assessment documents from government, independent institutions, and industry are examined in this chapter to determine the range of scientific opinions on the acid deposition problem and to document any pattern that might emerge between assessment findings and sponsorship of different reports. This comparison may assist policy formulation by identifying the extent of scientific agreement and disagreement on key issues.

Methodology

Two facets of the acid rain problem are examined: its physical and chemical characteristics, and its environmental effects. A matrix method of comparison is used, which is adapted from an earlier comparison of scientific assessments by Joan P. Baker.[1] A set of questions related to key aspects of the acid rain phenomenon is generated and then asked of each assessment. The responses are summarized in the text and presented in tabular form in appendixes A and B. The method allows for comparison of the major conclusions of the assessment documents.

This chapter was written by Kim J. DeRidder

Whenever possible, questions are designed to elicit an affirmative/ negative response. This reduces ambiguity in evaluating a given report's position. However, when qualifying statements are included in the document, it is so reported, and explanatory comments are offered; the document pages where each response is found are listed in the tables. Where a document fails to address a given topic, "not addressed" is entered for its response.

The Assessments

The assessment documents used in this chapter are listed in table 2-1. The choice of assessments subject to the full analysis is limited to those available prior to the publication of the first edition of this book. Two particularly significant reports, *Acid Deposition, Long-Term Trends*[2] and the interim report of the National Acid Rain Precipitation Assessment Program (NAPAP),[3] have been published since that time. We analyze these at the end of the chapter, together with a report entitled "Is There Scientific Consensus on Acid Rain?"[4] which analyzes six government (including NAS) reports, four of which were published in 1984.

Types of Assessments

Three types of documents are represented: (1) scientific assessments, (2) collections of individual studies, and (3) policy assessments. All three types are intended to assist policy formulation. Scientific assessments are produced by teams of scientists, either for government or private industry, and represent a synthesis and evaluation of current scientific research on the nature and significance of the acid deposition phenomenon. The second type of document is not integrated, and often consists of collections of individual studies that are bound into a single document. The third type of document, the policy assessment, represents the attempt to sift through available scientific literature to identify policy-relevant characteristics of the acid deposition problem. This type of document is more likely than the previous types to consider control options and their costs. Authors of the first two kinds of reports are often commissioned by government agencies or interest groups, while policy assessments are generally written by staff analysts working for government, industry, or independent groups.

Table 2-1.   Assessment Documents

I. Government Reports

OTA   "The Regional Implications of Transported Air Pollutants: An Assessment of Acidic Deposition and Ozone." Interim draft. Prepared by the Office of Technology Assessment, U.S. Congress, for the Senate Committee on Environment and Public Works and the House Committee on Energy and Commerce. Mimeo, July 1982.

MOI.1   *United States—Canada, Memorandum of Intent on Transboundary Air Pollution: Impact Assessment, Work Group I.* Final report. Washington, D.C., January 1983.
    This and all MOI documents used in this paper are joint U.S.-Canadian documents. In some cases Canada and the United States offer differing interpretations of the scientific data. This is done by including national position reports within the larger document. Because this is done frequently in MOI.1, the document can, to a small extent, be seen as two documents in one. MOI.1C and MOI.1US are used to identify the national position reports.

MOI.2   *United States—Canada, Memorandum of Intent on Transboundary Air Pollution: Atmospheric Sciences and Analysis, Work Group 2.* Final report. Washington, D.C., November 1982. (See explanation of MOI.1.)

MOI.3B   *United States—Canada, Memorandum of Intent on Transboundary Air Pollution: Emissions, Costs and Engineering Assessment, Work Group 3B.* Final report. Washington, D.C., June 1982. (See explanation of MOI.1.)

MOI.H   "Highlights from the Final Reports of the Canada/U.S. Work Groups Established under the Memorandum of Intent between the Government of Canada and the Government of the United States of America concerning Transboundary Air Pollutants." Mimeo [1983]. (See explanation of MOI.1.)

NAPAP   *National Acid Precipitation Assessment Plan.* Draft. Prepared by the Interagency Task Force on Acid Precipitation. Washington, D.C.: Council on Environmental Quality, 1981.

ARIB   *Acid Rain Information Book.* Prepared by Frank A. Record, David V. Bubenick, and Robert J. Kindya for the U.S. Department of Energy. Park Ridge, N.J.: Noyes Data Corporation, 1982.

CAD.1,   *Critical Assessment Document: The Acidic Deposition Phenomenon*
CAD.2   *and Its Effects,* vols. 1 and 2. Draft. Prepared for the U.S. Environmental Protection Agency through the North Carolina State University Acid Precipitation Program, October 1982. Washington, D.C.: U.S. Environmental Protection Agency, 1982.

Table 2-1.   (continued)

DYCO   *The Effects of Air Pollution and Acid Rain on Fish, Wildlife, and Their Habitats: Introduction*; and *The Effects of Air Pollution and Acid Rain on Fish, Wildlife, and Their Habitats: Results of Modeling Workshops.* Both prepared by M. A. Peterson, Dynamic Corporation, Rockville, Maryland, for the Eastern Energy and Land Use Team of the Office of Biological Services, Fish and Wildlife Service, U.S. Department of the Interior. FWS/OBS-80/40.3, 1982. Conducted as part of the Federal Interagency Energy Environment Research and Development Program of the U.S. Environmental Protection Agency.

NRCC   *Acidification in the Canadian Aquatic Environment: Scientific Criteria for Assessing the Effects of Acidic Deposition on Aquatic Ecosystems.* Prepared by the National Research Council Canada (NRCC), Associate Committee on Scientific Criteria for Environmental Quality, Subcommittee on Water. Publication NRCC #18475 of the Environmental Secretariat, 1981.

II. Independent Scientific Assessments*

NRC.1   *Atmosphere-Biosphere Interactions: Toward a Better Understanding of the Ecological Consequences of Fossil Fuel Combustion.* Prepared by the Committee on the Atmosphere and the Biosphere, National Research Council. Washington, D.C.: National Academy Press, 1981.

NRC.2   *Acid Deposition: Atmospheric Processes in Eastern North America.* Prepared by the Committee on Atmospheric Transport and Chemical Transformation in Acid Precipitation, National Research Council. Washington, D.C.: National Academy Press, 1981.

NCSU   "A Status Report on Acid Precipitation and Its Ecological Consequences as of April 1981." Prepared by Ellis B. Cowling, School for Forest Resources, North Carolina State University, as a result of interactions with the National Atmospheric Deposition Program, EPA/NAPD/NCSU Acid Precipitation Program, and the North Carolina Agricultural Research Service at NCSU. Mimeo, 1981.

III. Industry

EPRI   "Ecological Effects of Acid Precipitation." Prepared by the Central Electricity Research Laboratories for the Electric Power Research Institute (EPRI). Mimeo, workshop report, EPRI SOA77-403, July 1979. The workshop was held in England, on behalf of EPRI, an American firm.

EEI   *An Updated Perspective on Acid Rain.* Prepared by Alan W. Katzenstein for the EEI Acid Rain Public Response Task Force. Washington, D.C.: Edison Electric Institute (EEI), 1981.

Table 2-1.   (continued)

ON-HY   "Research Review: Acid Rain," no. 2, May 1981. A collection of ten individual research papers, prepared for the Ontario Hydro Company.

IV. Policy Assessments

LAW   *Acid Rain in Europe and North America: National Responses to an International Problem.* Prepared by Gregory Wetstone and Armin Rosencranz on behalf of the Environmental Law Institute as a study for the German Marshall Fund of the United States. Washington, D.C.: Environmental Law Institute, 1983.

SCAR   *Still Waters.* A report of the Subcommittee on Acid Rain, for the first session of the Thirty-second Parliament 1980–81 of Canada. Ottawa, Ontario: Minister of Supply and Services, 1981.

Note: Depending on subject matter covered, some assessment documents are used in both sections (physical and chemical characteristics, environmental effects). Others are used only once. The reader can check on the use made of each document in Appendixes A and B.
*Only U.S. assessments were available.

Selection Criteria

It would have been preferable to use only scientific assessments, as they are designed to guide policy formulation with an authoritative evaluation of available evidence. However, this was not always possible. In the section dealing with the physical and chemical characteristics of acid deposition a number of basic questions about the nature and severity of acid rain were not addressed by the assessments. Therefore, several collections of individual research findings and a policy assessment were included to broaden the base of the study.

In the section on environmental impact, two exceptions to the rule of concentrating on scientific assessments occur. A policy assessment by the Environmental Law Institute (LAW) was used, as well as a report by the Edison Electric Institute (EEI). LAW's report is included to serve as an example of current policy assessments on the subject. EEI's document is a brief report geared for a general readership. Because few industry assessments were available to the author, it is included to illustrate the scientific conclusions reached by the private sector.

Beyond these exceptions, the criteria for selecting reports included: (1) sponsorship by either government, independent organizations, or private industry;[5] (2) relative recency of publication; (3) focus on overall environmental impact (as opposed to impact on specific

Table 2-2.   Questions Concerning Aquatic Impact on Acid Deposition

1. Are lakes more acidic now than they used to be?

2. Can any increase in lake and stream acidity be related to an increase in atmospheric deposition of acids?

3. Is the distribution of acidified lakes and streams correlated with the distribution of acidic deposition?

4. Has atmospheric deposition of acids increased in recent decades coincident with the increased number of acid lakes and streams?

5. Are there other possible explanations for the regional increases in acidity of lakes and streams?

6. What is the extent of aquatic resources susceptible to acidic deposition in North America?

7. By how much would acidic deposition have to be reduced to prevent damage to sensitive aquatic ecosystems?

8. What damage to fish populations as a result of acidification has been documented?

9. What does each report conclude?

Source: Adapted from Joan P. Baker, "Acidic Deposition and Its Effects on Aquatic Ecosystems," unpublished report, Duke University, AAAS/EPA Environmental Science and Engineering Fellowship Program, 1982.

locales); and (4) Canadian or U.S. origin.[6] Initially the author searched for scientific differences of opinion between Canadian and U.S. documents, but on major issues no clear pattern resulted from this approach. This is a significant finding that will be discussed in greater detail later; it is mentioned here because of its importance for the selection of documents that were included in the review.

The Questions

Baker, in her previously mentioned study, compared statements concerning the aquatic impacts of acid rain in ten assessment documents. The questions she asked of the various documents are listed in table 2-2. This chapter goes further. In order to compare a broader range of topics, two new sets of questions were developed. The first set (see table 2-3) deals with the physical and chemical characteristics of the acid rain phenomenon. Specific questions address the historical and future trends in acidity and its precursors, anthropogenic sources of

Table 2-3.   Questions Concerning Properties of
Acid Deposition and Its Precursors

1.a. Does there appear to be a historical increase in the trend of quantity or intensity of acid deposition?

 b. Does there appear to be a historical increase in the trend of quantity or intensity of SO_2 emissions in North America?

 c. Does there appear to be a historical increase in the trend of quantity or intensity of NO_x emissions in North America?

2.a. What are the major sources of sulfur dioxide emissions in North America: electric utilities?

 b. What are the major sources of nitrogen oxide emissions in North America: electric utilities, transportation, and industry?

3.a. Have natural sources of nitrogen and sulfur compounds, though dominant sources globally, become less significant when compared with anthropogenic emissions on a regional scale in North America?

 b. Is damage resulting from dry acid deposition comparable to that of wet?

4. Can the precursors of acid deposition, or the resulting acids, be transported long distances from their sources (including to previously unaffected areas)?

5. Can damage to specific areas be attributed to point sources of emissions?

6. Is the relationship between emissions and acid deposition linear?

7. Are either transformation or transportation models accurate enough to be relied upon for development of control strategies?

8. Are increases in acid deposition or its precursors predicted for the future?

---

acid deposition precursors, the significance of human contributions to these precursors, and the predictability of precursors' transformation and transportation once they enter the atmosphere. The second set of questions (see table 2-5) addresses aquatic as well as terrestrial impacts of acid deposition, the primary causes of damage done to the environment, the permanence of damage, and whether remedial measures seem suitable.

Physical and Chemical Characteristics

This section begins with brief background information on the physical and chemical properties of acid deposition. It then discusses the significance of each of the questions in table 2-3. This is followed by a summary of the responses offered by the assessment documents. Specific document responses are tabulated in the appendixes to this chapter.

Background on Acid Deposition

Acids occur in both wet and dry forms. In solution, acidity is determined by the concentration of unattached positively charged hydrogen ions. (An ion is an electrically charged atom, and by nature hydrogen atoms are extremely reactive.) In highly concentrated solutions these ions possess the capability to break down many chemical compounds by displacing ingredients that have a somewhat weaker electrical charge. Acid rain or acid precipitation is caused by the removal of sulfates and nitrates from the atmosphere by rainfall, occurring in one of two ways. Atmospheric acid salts may be cleansed from the air beneath rain clouds by falling precipitation. Water droplets simply pick up the salts during their descent. Or, moisture in clouds can gather around salt (rather than dust) particles to form acidic raindrops or snowflakes.

Dry deposition includes several processes by which gases and particulate matter are deposited. These processes are not fully understood. Areas of uncertainty include: (1) how to measure rates of dry deposition; (2) how transport variables affect transformation of gases and aerosols; and (3) how characteristics of receptors themselves may influence the extent to which they attract deposition.[7] It is possible that the effects of dry deposition may be at least as destructive to the environment as acid precipitation. It may accumulate in dry particulate form on plants or structures and turn to acid as moisture becomes available (through fog, dew, transpiration, or other natural means). The resulting solution can be more concentrated and destructive than acid precipitation.

Significance of Questions and Summary of Document Responses

Historical Trends in Acid Deposition

1.a. Does there appear to be a historical increase in the trend of quantity or intensity of acid deposition?

This question is important as an indicator of the urgency of the problem and the need for a rapid response. If deposition is increasing, then any damages thus far associated with acid deposition can be expected to increase. If decreasing or stable, however, then perhaps it will be thought of as a less urgent matter, with more time available for policy development.

Studies of trends in North American acid deposition have been the subject of considerable debate, centering on the accuracy or comprehensiveness of historical data. While many researchers have claimed that rainfall is becoming more acidic over broad areas, others point to differences in collection methods, siting criteria, chemical analysis techniques, quality assurance measures, and methods for the storage of samples to establish that available data for estimates are not conclusive.

The general consensus of the assessment documents is that no statement can be made as to whether the environment has experienced increasing or decreasing trends in acidic deposition (see appendix A, table A-1). The private industry document goes further and states specifically that no trend exists.

1.b. Does there appear to be a historical increase in the trend of quantity or intensity of SO_2 emissions in North America?

1.c. Does there appear to be a historical increase in the trend of quantity or intensity of NO_x emissions in North America?

Even if a verifiable trend in the increase of acid deposition cannot be determined, trends in the precursors of acid deposition can serve as an alternative predictor. Increases in precursor emissions would suggest that acidity will increase, though the ratio of acidity deposited to emissions released would depend on the nature of the transformation process (see question 5).

Almost all the reports concluded that emissions of both substances have shown a historical increase, but many reports added qualifications to this statement. In regard to SO_2, an industry-sponsored assessment reported that increases had occurred in the United States but could not be determined for Canada. In two different volumes of the same document (MOI.2 and MOI.H.) the responses are inconsistent, the first stating that trends could not be determined and the second that they could. One contracted agency report stated specifically that there were no increased trends for SO_2.

In regard to NO_x, six of seven assessments stated that trends show increasing amounts in North America, again with the private industry document indicating this was true for the United States but could not be determined for Canada. Only the MOI document said a trend could not be determined for North America as a whole. Tables A-2 and A-3 summarize the responses.

Sources

2. What are the major sources of acid deposition precursors in North America?

The combustion of fossil fuels involves large volumes of raw materials, occurs at high temperatures, and is a major source of anthropogenic substances in the atmosphere.[8] Emissions from burning fossil fuels ordinarily enter the atmosphere as gases (e.g., nitrogen oxides, sulfur dioxides). They can also enter as solid or liquid particles, referred to as primary aerosols.

In an effort to establish a definite relationship between acid deposition and its precursors, it is useful to assess the amount of emissions currently being released into the atmosphere. Emissions data, usually calculated from the amount of different fuels consumed, aid in predicting acid deposition trends.

Sulfur oxides are generally agreed to come primarily from electric utilities. Specifically, seven documents indicated that electric utilities are responsible for the majority of SO_2 emissions in North America, although two (MOI and ON-HY) added a qualifier that this is true for the United States but not for Canada. One document, DYCO, stated that the primary sources are almost equally stationary fuel combustion and industrial processes.

Nitrogen oxides are emitted by a variety of sources. Five documents indicated that the major sources of NO_x in North America are (roughly) equally electric utilities, transportation, and industry, with again two (MOI and ON-HY) adding that this is true for the United States but not Canada. Two contracted U.S. government assessments stated that the primary sources are either electric utilities or stationary sources (inclusive of electric utilities) and transportation. Tables A-4 and A-5 summarize the responses.

Anthropogenic Versus Natural Precursors

3.a. Have natural sources of nitrogen and sulfur compounds, though dominant sources globally, become less significant when compared with anthropogenic emissions on a regional scale in North America?

Although many, if not most, scientific assessments indicate that the largest portion of acid deposition precursors, on a worldwide scale, is natural in origin, they also point out that human contributions downwind of urban centers and large point sources are highly significant. In some areas precursor loadings may have reached levels high enough that even small additional increases in emissions may have significant impacts. Moreover, assessments recognize the importance of better understanding the relative proportions of natural and man-made depositions at the regional level, particularly when control measures are being discussed. The importance of anthropogenic emissions on a regional scale is universally agreed to. Of the eight documents addressing this question, all state that man-made sources of the precursors to acid deposition are responsible for the effects of acid deposition. Table A-6 presents the responses to this question.

Dry Versus Wet Deposition

3.b. Is damage resulting from dry acid deposition comparable to that of wet?

Some studies suggest that, because moisture plays a critical role in the transformation process, and because moisture is relatively more available at the ground level, unconverted deposited SO_2 may oxidize to sulfate very quickly, perhaps making ground-level conversion a more important problem than its atmospheric counterpart. If true, the amount of sulfur deposited in any form may be the most crucial consideration for environmental impact, and NO_x may be less significant. In addition, little is known about dry deposition because most attention has focused on wet deposition (hence the name acid rain). The magnitude and severity of effects of dry deposition are critical factors in assessing control options and their expected benefits.

Responses to this question were split between positive and uncertain. Five documents stated that dry acid deposition is comparable in quantity to wet. One U.S. contracted assessment said the amount could not be determined. Only the Canadian policy assessment said it was less than wet. The results suggest that dry deposition is at least a potentially serious problem and is most likely a real threat. As such, it needs to be taken into account in the solution of the acid rain problem. Table A-7 presents the assessment results.

Long-distance Transport

4. Can the precursors of acid deposition, or the resulting acids, be transported long distances from their sources (including to previously unaffected areas)?

What becomes of a pollutant once it is emitted into the atmosphere is contingent upon many factors, some meteorological, others a characteristic of the pollutant. Weather and global wind patterns strongly influence the transport process. A pollutant's residence time also plays an important role.[9] The height from which emissions are released can also have a significant bearing on the distance of transport. One of the pressing issues concerning transport has to do with the likelihood of acids, or their precursors SO_2 and NO_x, being carried long distances from the point of emission to the point of deposition. There are currently reports of acid deposition recorded in remote and nonindustrialized areas. If acid deposition can be transported long distances, emitters of acid deposition precursors, especially large emitters like utilities, may be responsible for environmental damage to areas far beyond state or national boundaries. All documents stated that pollutants causing acid deposition can be transported long distances (see table A-8).

Relating Specific Emitters to Areas of Damage

5. Can damage to specific areas be attributed to point sources of emissions?

Knowing whether utilities and other emitters are responsible for acidity deposited in areas located great distances from the site of combustion is important for implementing and financing controls. Determining appropriate control strategies will be difficult if damage to specific areas cannot be tied to its origin. If the emitters responsible for damages to a site were known, control strategies tailored for those sites could be established, thus avoiding the imposition of excessive controls on other sources. The eight documents addressing this question all agreed that tracing environmental damages to point sources of emissions is not possible with presently available information. Table A-9 details the responses.

It is generally understood that acid deposition is the result of a transformation of ambient sulfur dioxide and oxides of nitrogen into sulfuric acid, nitric acid, the anions sulfate and nitrate, and the cation ammonium. This process involves the addition of oxygen to the gaseous precursors. Several chemical pathways through which transformation may take place have been identified. The most straightforward route is reaction of sulfur and nitrogen pollutant gases with oxygen gas. However, evidence suggests that more complex reaction pathways involving oxidant pollutants, such as hydrogen peroxide and ozone, water vapor, and sunlight, play a more significant role in the formation of sulfates, nitrates, and acids. One study indicates that the most important reaction pathway involves oxidant pollutants and tiny atmospheric water droplets known as aerosols. However, it is generally felt that "(t)he transformation of sulfur dioxide to sulfate aerosol and/or sulfuric acid in the atmosphere is a key aspect of the acid deposition problem."[10]

Relationship between Emissions and Deposition

6. Is the relationship between emissions and acid deposition linear?

Only four of the nine documents addressed this question. Of the four, two (NAPAP and DYCO) said the quantitative relationship between precursors and acid deposition could not be determined. One (MOI) said the relation was nonlinear over short distances and undetermined over long distances. Only NRC.1 stated the relationship to be linear.

Some studies suggest that acid deposition has been worsening in recent decades not only because of an increase in the emissions of acid precursors, but also due to an increase of oxidants that make the atmosphere more reactive. (Oxidants, like acid deposition, are not emitted directly; they result from photochemical reactions between nitrogen oxides and hydrocarbons.) If true, this would suggest that the relationship between precursor inputs and acid deposition outputs could be nonlinear. A given reduction of SO_2 emissions will have different effects on the amount of acid deposition depending on the shape of this relationship. How this issue will be settled has important consequences for the selection of appropriate control measures, and for the costs of achieving a given reduction. Several assessments indicated that the question is unresolved; other assessments did not address the question at all. The responses are presented in table A-10.

Adequacy of Models

7. Are either transformation or transportation models accurate enough to be relied upon for development of control strategies?

While transformation as well as transportation models have become better in their ability to predict the physical and chemical pathways of acid deposition and its precursors, it is important to know just how accurate they are. If they are reliable, they can be used now for policy development: with predictable input/output relationships, more efficient and effective control strategies can be determined. If available models are not reliable, less exact solutions, with increased risk of both over-control and under-control, will result.

Four U.S. government documents stated that neither existing transformation nor transportation models are accurate enough to be relied upon for policy-making. Only the private industry document expressed confidence that models can guide action. MOI suggested that models might possibly be used. The prevailing opinion of the responding assessments was that as yet neither type of model was sufficiently accurate to guide policy development (see table A-11).

Future Trends

8. Are increases in acid deposition or its precursors predicted for the future?

If acid deposition can be expected to level off or even to decrease relative to current levels, the urgency of taking remedial action would be lessened. If significant increases are predicted, immediate imposition of control measures becomes more important.

All six of the documents addressing this question stated that increases in acid deposition and/or its precursors could be expected for the future in North America (see table A-12).

U.S. versus Canadian Contributions to Acid Deposition

Unfortunately, an important question concerning the nature of the acid rain problem in North America could not be addressed. The author had planned to include a question on the international aspect of the acid rain phenomenon, but found that the question was not addressed in most assessments.

Regarding cross-border contributions, the question to be asked is: What amount of the United States' deposition is carried to Canada, and vice versa? If indeed acid deposition and its precursors are able to be transported long distances, then, given the observation that summer wind currents generally travel from south to north in eastern North America,[11] emitters from the United States may be responsible for portions of Canada's acid deposition. Likewise, North American winter breezes tend to flow south, suggesting Canada may likewise be responsible for some portion of acid deposition in the United States. The additional observation that the atmosphere tends to be more reactive (able to convert more emissions to acid deposition)[12] in the summer compounds the effects of the possible U.S. contribution to Canada. The only assessment to address the question was SCAR, which concluded: "Preliminary estimates suggest that about 3 to 4 times as much sulfur, on an annual average basis, moves across the border from the United States to Canada than moves in the opposite direction. A reasonable estimate of the U.S. contribution to Canada's sulfur-sourced acid rain is approximately one-half, with the other half originating from domestic emissions; in some sensitive areas ... the United States contribution is as high as 70 percent."[13] Because the statement is not referenced to specific sources, and since later documents do not confirm this information, the results of additional monitoring programs will be needed before the question can be answered. Confirmed results, obviously, will be important for guiding a bilateral control program.

Range of Agreement and Disagreement

There is complete agreement among assessment documents that man-made sources of the precursors to acid deposition are responsible for the observed effects of acid deposition; that acid deposition and its precursors can be transported long distances through the atmosphere; that damage to specific areas cannot be traced to specific sources of the precursor emissions; and that increases in either acid deposition and/or its precursors can be expected for the future.

Table 2-4.   Assessment Responses on Physical and Chemical Characteristics

Docu-	Question 1			Question 2		Question 3	
ment	a	b	c	a	b	a	b
MOI.2	DK	DK	DK	Y/N			Y
MOI.3B					Y/N		
MOI.H		Y				Y	
NAPAP	DK	NA	NA	Y	Y	Y	Y
CAD.1	DK	Y	Y	Y	N	Y	NA
DYCO	NA	N	Y	N	N	Y	DK
NRCC	DK	Y	Y	Y	Y	Y	Y
NRC.1	NA	Y	Y	Y	Y	Y	NA
NRC.2	DK	NA	NA	NA	NA	Y	Y
OH-HY	N	Y/DK	Y/DK	Y/N	Y/N	Y	Y
SCAR	DK	Y	Y	Y	Y	NA	N

Response abbreviations:
Y   =   Yes
N   =   No
M   =   Maybe
DK =   Don't know
NA =   Not addressed

There is nearly complete agreement that there has been a historical increase in the trend of man-made emissions of NO_x; that the major sources of man-made SO_2 emissions in North American are electric utilities; and that no statement can be made as to whether the environment has historically experienced increasing or decreasing trends in acidic deposition.

There is somewhat less agreement that the major sources of NO_x in North America are electric utilities, transportation, and industry, each contributing roughly one third; that the amount of dry acid deposition in North America is comparable to that of wet; and that neither transformation nor transportation models are accurate enough to be relied upon for development of refined control strategies.

There is least agreement that there has been a historical trend in increased SO_2 emissions in North America; and that the relationship between emissions and resulting acid deposition is linear.

Assessment responses to the eight questions relative to the physical and chemical characteristics of acid deposition are tabulated in table 2-4.

Docu-ment	Question				
	4	5	6	7	8
MOI.2	Y	N	N/DK	M	
MOI.3B					Y
NAPAP	Y	N	DK	N	Y
CAD.1	Y	N	NA	N	NA
DYCO	Y	N	DK	N	Y
NRCC	Y	NA	NA	NA	Y
NRC.1	Y	N	NA	NA	Y
NRC.2	Y	N	Y	N	NA
OH-HY	Y	N	NA	Y	NA
SCAR	Y	N	NA	NA	Y

Environmental Effects

In this section we examine the second set of questions, those relative to the environmental impacts of acid rain (see table 2-5). The significance of each question is discussed first, followed by a summary of the responses offered by the assessment documents. Detailed document responses are tabulated in appendix B.

Significance of Questions and Summary of Document Responses

Forms of Environmental Damage

1. Does acidification substantially damage aquatic ecosystems?

2. Does acid deposition substantially damage terrestrial vegetation?

3. Does acid deposition substantially damage soils?

Environmental impact occurs as damage to aquatic and terrestrial ecosystems and to soils. "Aquatic ecosystem" is defined here as a self-

Table 2-5.   Questions Concerning Environmental Impact of Acid Deposition

1. Does acidification substantially damage aquatic ecosystems?

2. Does acid deposition substantially damage terrestrial vegetation?

3. Does acid deposition substantially damage soils?

4. What are the main causes of damage to the aquatic ecosystem with regard to acid deposition?

5. Given a decrease or termination of acid deposition, would the incurred adverse effects reverse naturally with time?

6. For acidic aquatic ecosystems, does liming appear to be a feasible remedy?

7. Does the impact of acid shock as a product of snowmelts result in more damage to aquatic chemistry and/or ecosystems than if the same impact had been gradual?

sustaining and self-regulating community of organisms interacting with one another and with their aquatic environment.[14] Baker's comparison of ten acid rain assessments found agreement that "acidic deposition is causing acidification of surface waters and damage to fish populations."[15] The question asked in this report is whether acidification causes serious problems not only to fish but to the entire aquatic ecosystem.

Adverse impact can take different forms, including imbalance, alteration, degeneration, or death of some or all members of the independent food and life-ways web of one or more species of organisms within an ecosystem. Damage from acidification of the aquatic ecosystem is often defined as degeneration, reduction, or extinction of species populations of fish, reptiles, crustaceans, microbiotic life, insects, and aquatic vegetation. Substantial damage can be seen as degeneration, reduction, or extinction of any key member(s) of the ecosystem.

In general, the assessments surveyed are in agreement that acidification can and often does degenerate, reduce, or destroy biotic species populations and damage certain aquatic ecosystems. One U.S. government document simply states that a correlation exists. The private industry documents are less conclusive.

The effect of acid deposition on forest ecosystems is often broken down into three areas of causal agents: acidity, sulfur and nitrogen, and ozone. Sometimes there are discussions of synergistic effects resulting from acidity changes and ozone.

There is agreement among researchers that, while several of the impacts on forests may be adverse, nitrogen deposition within certain ranges may produce beneficial effects.[16] Consequently, our survey of assessment documents searches for the net effects of acid deposition on forest environments, generally in terms of degeneration of species populations of various plants, or their health. Health may be construed in terms of productivity, or the ability of the plant to grow and/or reproduce.

While acid-induced changes in forest soil are believed to have important forestry impacts, most reports address the topic of soils separately.[17] The greatest concern is that aluminum ions displaced from aluminum silicates by acid precipitation will appear in the soil and then be leached into aquatic ecosystems. Concentrations of aluminum have been demonstrated to have adverse effects on terrestrial vegetation as well as on aquatic biota.[18] For present purposes, "soil damage" is defined as radical alteration of the physical and/or chemical makeup of soil.

Of the three areas of environmental impact—aquatic, terrestrial, and soils—only damage to the former has been extensively documented in the field. This is due in part to the fact that organisms living in water (which dissolves and disperses chemical inputs throughout the ecosystem) must at all times live in direct contact with new chemicals or chemical imbalances. Consequently, organisms in an aquatic environment will likely be the first members of the biota to show signs of impact. In addition, aquatic environments are the final stops in watersheds. Acid deposition that falls on land either washes into waterways or interacts with soils. As a result, toxic metals may wash into waterways. Hence, aquatic environments generally incur greater impacts relative to their terrestrial counterparts. Knowledge of impact on soils and terrestrial vegetation, due to this lack of field documentation, so far is mostly derived from laboratory experimentation.

Assessments nearly consistently report that impacts on terrestrial vegetation are, to date, largely unknown, despite the fact that a great deal of research has been performed on the subject. One independent assessment states that damage occurs from dry deposition of toxic gases. Recent reports on damage to forests were not available at the time the assessment documents included in this comparison were prepared.

Almost every type of assessment represented in the review indicates that acid deposition can and often does substantially damage soils. Only private industry is noncommittal on the subject.

Tables B-1, B-2, and B-3 tabulate the document responses.

Causes of Damage to the Aquatic Ecosystem

4. What are the main causes of damage to the aquatic ecosystem with regard to acid deposition?

There is agreement that acid deposition does have adverse impact upon aquatic environments. We ask whether there is also agreement among reports as to the causes of that impact, and what is it about acid deposition that causes the damage? If agreement can be found to exist, perhaps policies can be created that specifically bar, limit, or divert industrial waste inputs that contribute to these causes.

Almost across the board, assessments agree that damage to aquatic ecosystems is strongly correlated with pH depressions and the effects of toxic metals leached from nearby soils. All assessments except for two industry reports agree on these two factors. Some documents suggest other possible agents. (See table B-4 for the detailed responses.)

Reversibility and Postpositional Remedies

5. Given a decrease or termination of acid deposition, would the incurred adverse effects reverse naturally, with time?

6. For acidic aquatic ecosystems, does liming appear to be a feasible remedy?

Given agreement that there is an impact on aquatic environments, it is important to ask whether effects can be reversed, either naturally, following termination of acid deposition, or by intervention with chemicals. The issue of reversibility both addresses the permanence of the effects of acid deposition and suggests strategies for possible policy responses.

Most documents reviewed do not address the topic of natural reversibility of adverse effects of deposition on aquatic ecosystems following reduction or elimination of acid inputs. Those that do indicate that adverse impacts are only partially reversible. Of the assessments addressing the question of reversibility, less than half concluded that damages might be at least partially reversed through natural processes. In documenting the responses of assessments, estimates about time frames of natural reversibility are provided in the "comments" section of table B-5.

Different ways of assisting the rejuvenation of aquatic environments have been considered. Specifically, liming has received much attention and is considered by some to be the most promising form of corrective action. Liming is the introduction of alkaline lime into aquatic ecosystems to negate the effects of acidity and to restore more normal pH levels. Therefore, assessments were examined to determine the extent that liming is considered to be a viable response.

The assessments that did address this question generally indicated liming might be a partial solution, but they also qualified their responses by pointing out the uncertain effects and high costs of an extensive program. These qualifications cast doubt on the use of liming as a long-term solution. In terms of institutional authorship interesting differences emerge: most U.S. government documents (in-house as well as contracted) view liming as sometimes useful. Private industry reports vary from yes to no. Independent research and Canadian government documents state that liming is not a promising response. Table B-6 summarizes the responses.

Acid Shock and Gradual Impacts

7. Does the impact of acid shock as a product of snowmelts result in more damage to aquatic chemistry and/or ecosystems than if the same impact had been gradual?

Two patterns of disruption appear as a result of sudden or gradual variation in acid input. Sudden surges in acid deposition occur when snow and ice, having collected deposition through the winter, thaw with spring temperatures, causing sudden "shocks" of acid to aquatic systems. Often shocks are sufficiently severe to result in fish kills at pH levels above those normally toxic to fish.[19] Gradual acidification is the result of direct input of acid deposition through wet and dry deposition processes, in addition to deposition that results from water runoff from the land. Field observations and laboratory experimentation indicate that prolonged acidity interferes with fish reproduction and spawning such that, over time, there is a lowering in population density of fish. There is also a shift in the age and size of the population to older and larger fish, as younger fish tend to be more affected by increased acidity.

Although both gradual and sudden means of disruption can produce fatal results, it seems that sudden shocks can have a greater net adverse impact on ecosystems than gradual acidification, per unit input. If true, Canada would be more vulnerable to acid deposition as a function of its northerly climate. For this reason the question is important in searching for a bilateral solution.

It must be noted that the assessments surveyed seldom specifically compare the extent to which sudden surges are more detrimental than gradual input. They do, however, usually address each issue separately. Moreover, documents tend to reserve stronger language for describing impact due to acid shock than to gradual impact. Therefore, while documents may not specifically state that the former is more detrimental per unit input than the latter, this relationship is implicit in the wording.

Another conclusion drawn from responses to this question has to do with the use of the word "significant." A document simply stating that the impact of acid shock is significant seems also to carry the implication that it is more detrimental per unit impact than gradual input. It has been demonstrated that months' worth of deposition is stored in ice and snow, perhaps even collecting in such a way that it radiates to the outermost regions of the ice formation. Rapid thawing of that ice and snow will result in massive quantities of acids being introduced into the aquatic environment. If very gradual inputs are significant, then of course sudden massive inputs will be significant. If gradual small doses of acids have a significant impact, the type and degree of impact that will result from a sudden massive dose would virtually have to be greater.

Almost every document surveyed indicates that the impact on aquatic ecosystems of acid shock due to spring snowmelts is significant. Per quantity of acid input, it is suggested to have more detrimental effects on biota than gradual inputs. Table B-7 provides the summary of responses. Italics are used to indicate statements that seem to support the author's interpretations.

Range of Agreement and Disagreement

Baker, in surveying the ten documents used in her review, had found agreement on the following points:[20]

Acidic deposition is causing acidification of surface waters and damage to fish populations.

Quantification of the extent of the current damage and prediction of future damage is not possible with present state of knowledge.

Mechanisms of how lakes become acidified, why some lakes and fish populations are more susceptible than others, and functional relationships between atmospheric acid inputs and chemical and biological responses need to be better understood.

Initial approximations reflect large degrees of uncertainty because of inadequate understanding of mechanisms of response.

Our comparison further documents *almost complete agreement* among assessments that acid deposition is seriously affecting some of the aquatic ecosystems of North America; that the primary factors associated with this damage are depressions in pH and the presence of toxic metals leached into aquatic systems; and that variations in hydrology, especially acid shock due to springtime snowmelts, may have more adverse impact per quantity of acid input than does gradual acidification.

Several of the assessments surveyed indicate that acid deposition has serious effects on terrestrial soils.

We also document that there is *considerable disagreement* on whether deposition seriously affects terrestrial ecosystems, whether liming (the introduction of alkalines) is a feasible response to the problem of acidified aquatic ecosystems, and whether a few or many aquatic systems will be or have been affected.

Other Questions Considered   Although the questions examined so far cover the most important environmental impacts, other issues require attention. Three additional questions the author had originally included in the comparison had to be excluded because they were not addressed in enough assessments to allow for meaningful comparisons.

One question to be used was, "What would be an optimum pH for precipitation or for lakes and streams in order to protect aquatic ecosystems?" or, similarly, "What is the optimum allowable dose-effect acid loading that would still protect the environment?" Most reports failed to produce specific figures, but table B-8 summarizes the available responses to these questions.

Another question that was considered was, "What will be the trend of environmental impact of acid deposition in the future?" Table B-9 summarizes the responses. A third question that would have been useful is, "What is the relative sensitivity of both aquatic and terrestrial ecosystems in North America to acid deposition?" Unfortunately, documents used different measuring, quantifying, and qualifying techniques, making comparisons impossible.

In addition to statements about the effects of acid deposition and liming on aquatic environments, the MOI.1 document also addressed effects on terrestrial wildlife and ecosystems. It indicated that "direct effects of acid deposition on terrestrial wildlife have not been reported and are not considered likely. Nevertheless, indirect effects have been suggested," including reduction or losses in food sources, impairment of habitats, and contamination by heavy metals mobilized by acidity (pp. 1-14). The report also indicated that "the liming of forest lands to neutralize potential acidic deposition effects on terrestrial ecosystems has serious limitations" (pp. 1-24). The CAD.2 document suggests alternative ameliorative strategies to aquatic liming, including selective breeding, genetic screening, and acclimation (Chap. E-5, pp. 204–8).

Impact of Sponsorship of Assessments

The assessments that have been used in our survey were prepared by scientists and other experts working for different organizations —public and private, scientific and administrative, American and Canadian. How do the analysts associated with these institutions view the seriousness of the acid deposition problem? Can we detect a correlation between sponsorship and seriousness/urgency? If so, how can it be explained?

In order to respond to these questions, the queries and responses of the section on environmental impact are first presented in a measurable fashion by placing responses on a seriousness continuum, with *very serious* and *not very serious* as its poles. In most cases the affirmative response represents the most extreme position an assessment can take on the seriousness of a given situation. For example, to the question "Does acid deposition substantially damage aquatic ecosystems," *yes,* as opposed to *probably, maybe, probably not,* or *no* suggests that acid deposition poses a serious threat to aquatic ecosystems. (Questions 1, 2, 3, and 7 are of this type.) In other cases, as with the question "Given decrease or termination of acid deposition, would its effects reverse," the opposite is true, and *no* is the response that presents the most serious scenario. (Questions 5 and 6 are of this type.)

Question 4, "What are the main causes of damage by acid deposition to the aquatic environment . . . " essentially qualifies an assessment's response to question 1. Therefore, question 4 is not included.

A value system is attached to this continuum such that the values between + 2 and − 2, inclusive, are distributed between *very serious* and *not very serious* responses. Because anything between *yes* and *no* is subject to interpretation, the reader may want to compare the author's judgement with the information on which it is based (see tables in Appendix B). The scoring scale used is presented in table 2-6.

Table 2-6.   Values Assigned to Environmental Impact Responses

Question number	Value	Response
1, 2, 3, 7	2	Yes
	1	Probably, likely, partly, partially, sometimes
	0	Maybe, don't know, possibly
	−1	Probably not, not likely, doubtful
	−2	No
5, 6	2	No
	1	Probably not, not likely, doubtful
	0	Maybe, don't know, possibly
	−1	Probably, likely, partly, partially, sometimes
	−2	Yes

The values of responses are then summed for each document to produce a raw score. An assessment document depicting the most critical scenario, given that it responded to all seven questions, would have a raw score of 14. Similarly, a document portraying acid deposition as no real threat to the environment, having answered all questions, would receive a raw score of − 14. However, most documents failed to address every topic. Therefore, for each document the raw score is divided by the absolute value of the number of potential points, corresponding to the number of questions it addressed. For instance, if a document addressed all seven questions, the denominator under the raw score would be 14; if six, then 12; if four, then 8; etc. Consequently, final scores range in extremity from + 1 (very serious) to − 1 (not very serious). The resulting scores are presented in table 2-7.

Table 2-8 presents the range of assessment responses by institutional authorship of the assessment documents.

Policy Implications

This chapter has assessed the information available to policymakers on a number of scientific questions about acid deposition. How does the nonexpert policymaker respond to these comparisons? What conclusions can be drawn that can facilitate policy formulation? Perceptions of three factors are likely to shape the answers to these two questions: the seriousness of the problem; the uniqueness of the problem; and the existence of a number of areas of agreement and disagreement.

Table 2-7.   Assessment Responses on
Environmental Impact and Their Values

Docu-ment	Q 1 r/v	Q 2 r/v	Q 3 r/v	Q 5 r/v	Q 6 r/v	Q 7 r/v	R	Response quotient
OTA	Y/2	DK/0	Y/2	Pt/−1	NA	Y/2	5/10	.50
MOI.1C	Y/2	L/1	DK/0	Pt/−1	S/−1	Y/2	3/12	.25
MOI.1US	Pr/1	L/1	DK/0	Pt/−1	S/−1	Y/2	2/12	.17
NAPAP	Y/2	DK/0	NA	NA	NA	Y/2	4/6	.67
ARIB	Y/2	DK/0	DK/0	NA	Ps/0	M/0	2/10	.20
CAD.2	Y/2	DK/0	L/1	NA	S/−1	Y/2	4/10	.40
DYCO	Y/2	Pr/1	Y/2	NA	NA	Y/2	7/8	.88
NRCC	Y/2	NA	NA	Ps/0	Ps/0	Y/2	4/8	.50
NRC.1	Y/2	DK/0	Y/2	NA	DK/0	Y/2	6/10	.60
NCSU	Y/2	Y/2	Y/2	NA	NA/−	Y/2	8/8	1.00
EPRI	DK/0	DK/0	DK/0	NA	N/2	NA	2/8	.25
EEI	M/0	DK/0	NA	NA	Y/−2	Y/2	0/8	.00
LAW	Y/2	L/1	Pr/1	PN/1	N/2	Y/2	9/12	.75

Heading abbreviations:

Q	= question	R	= ratio of raw score to
r/v	= document response and its prescribed value		potential raw score

Response abbreviations:

Y	= Yes	PN	= Probably not
Pr	= Probably	NL	= Not likely
S	= Sometimes	N	= No
Pt	= Partly	G	= Given
L	= Likely	NG	= Not given
M	= Maybe	NA	= Not addressed
DK	= Don't know		
Ps	= Possibly		

Seriousness

As the results of this review point out, there is consensus among all assessments that acidification is substantially damaging some aquatic ecosystems in North America, and that this damage is, at best, only partially reversible. In jeopardy are not only the future of plants and animals directly or indirectly dependent upon these aquatic ecosystems, but also any economic or recreational activities that rely on these resources. Since soils are seriously threatened as well, forest ecosystems and their human and nonhuman dependents may be similarly affected.

Although past trends of acidification are inconclusive, in large part

Table 2-8.   Range of Assessments of Environmental Impact

Assessment Group	Range of assessment responses within groupings	
	Range	Average
U.S. in-house	.17 to .67	.45
U.S. contracted	.20 to .88	.49
Canada in-house	.25 to .50	.38
Independent	.60 to 1.00	.80
Industry	.00 to .25	.13
All assessments	.00 to 1.00	.48

perhaps due to inadequate data, acid deposition's precursors are expected to increase in the future. Thus in all likelihood the acid rain problem will worsen. The general agreement among assessments about the seriousness of acid rain means that the problem cannot be ignored by policymakers. The likely increase in acid deposition precursor emissions will require some form of new controls.

Table 2-9 ranks the assessment documents by the degree of seriousness they assign to the problem. Figures 2-1, 2-2, and 2-3 show how perceived seriousness of the acid rain problem differs among the institutions that have been involved in the assessment process. The most important finding, perhaps, is the absence of a significant difference in perceived seriousness between U.S. and Canadian assessment documents. Industry sees the problem as less serious, and independent and policy reports see it as more serious than government-authored or government-contracted documents.

Uniqueness

There is general agreement among most of the assessments on the factors that make acid rain different from previous environmental policy issues. Long-range transport and transformation are accepted by all the assessments, as is a lack of ability to trace specific damages to specific sources. The knowledge that we have today and the areas of agreement which exist now, all point to acid rain being different from previous phenomena, and therefore to a need to develop new types of policy for the solution. As always, designing new policy and finding a sufficiently large consensus for its implementation are hardest when past experience is of little help in guiding action. Acid rain, and perhaps global climatic changes induced by human activities, represent

Table 2-9.   Assessments of Environmental Impact in
Descending Order of Perceived Seriousness

	Document	Response quotient	Year	Sponsor	Score
Very serious (1.00)	NCSU	1.00	1981	U.S. government	1.00
•	DYCO	.88	1982	U.S. government	.88
•	LAW	.75	1983	Independent	.75
•	NAPAP	.67	1981	U.S. government	.67
•	NRC.1	.60	1981	Independent	.60
•	NRCC	.50	1981	Independent/ Canadian	.50
•	OTA	.50	1982	U.S. government	.50
•	CAD.2	.40	1982	U.S. government	.40
•	MOI.1C	.25	1983	Bilateral	.25
•	EPRI	.25	1979	Industry	.25
•	ARIB	.20	1982	U.S. government	.20
•	MOI.1US	.17	1983	Bilateral	.17
Neutral (0.0)	EEI	.00	n.d.	Industry	.00
•					
Not very serious (−1.00)					

new classes of pollutants that affect large areas of the globe and may cause irreversible damage. Given current knowledge about its nature and effects, acid rain seems to be the first of the new class of pollutants that cannot wait much longer for an appropriate political solution.

Areas of Agreement

There is widespread agreement among scientists on a number of major questions concerning the properties and effects of acid deposition. There is agreement that: the problem is worsening; anthropogenic sources are responsible for a large part of the problem; the primary sources involved are electric utilities, the transportation industry, and other industrial sources; the problem is severe and likely to worsen; long range transportation and transformation of substances can and does take place; and on a number of other questions. This emerging consensus is not strong enough or broad enough to support a solution to the problem of acid rain. Since acid rain has only been a major issue for ten years or less, the existence of any degree of consensus is a hopeful sign for the future resolution of the issue.

The areas of agreement are broad enough to support the development of a policy, however, and the structuring of the search for a

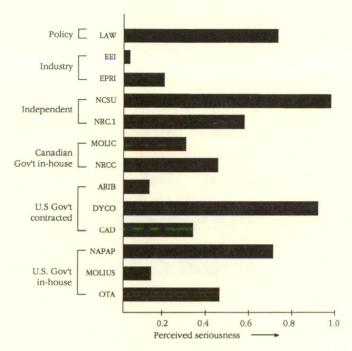

Figure 2-1.   Assessment Responses by Institutional Affiliation

solution. Transport and transformation models, research into effects
on soils and vegetation, and the like are important areas where there
is less certain knowledge. Commitment of research funds to those
areas would be one way to speed resolution of the problem. We do
know enough about acid rain now to decide where we should go to
learn more—what questions need to be asked and where to look for
the answers.

What is most clear is that we do not know enough to resolve the
acid rain issue with science alone. Nor are we likely to develop a
ready-made solution through more research. Given the consensus
that the problem is worsening, development of an acid rain policy is
an urgent task. The new policy will not be able to focus on control of
individual sources or setting of ambient air quality standards, the two
traditional tools for air pollution control, because of the regional scale
of the problem and lack of ability to tie damages to specific emissions.
Thus whatever policy emerges will have to be based on a different
policy model. Development of such a new policy will take time and
resources and should include a research plan to resolve as many of

Figure 2-2.   Assessment Responses by Institutional Groupings

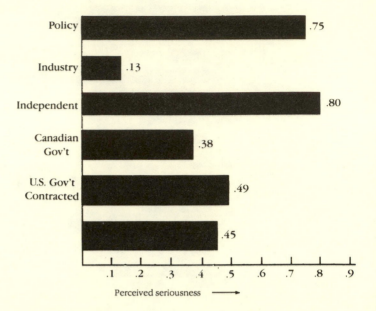

the open scientific questions as possible. Significant policy decisions will have to be made, however, before complete information is available, if indeed it ever is. Action must be taken before irreversible damage has been done.

More Recent Assessments

The National Academy of Sciences has published several reports in recent years.[21] The NAS has consistently linked acid deposition to anthropogenic emissions associated with highly industrialized and urbanized areas. In its most recent report, *Acid Deposition: Long-Term Trends*, NAS found a strong association between five parameters: SO_2 emission densities, concentrations of sulfate aerosols, visibility, sulfate concentrations in wet deposition, and sulfate fluxes in U.S. Geological Survey benchmark streams. They found that substantial quantities of SO_2 have been emitted over eastern North America since the early 1900s, fluctuating near current amounts since 1920. Since 1970 the southeastern United States has experienced the greatest rates of increase. Changes in sulfate observed in streams are consistent with

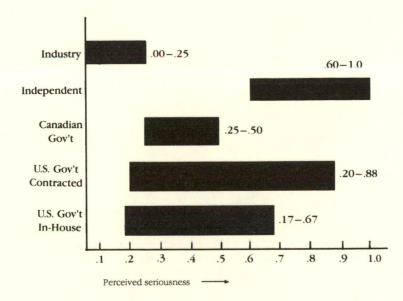

Figure 2-3.   Range of Assessment Responses by Institutional Groupings.
Note: Because only one policy assessment was reviewed, the policy
category is not included here.

changes in SO_2 emissions on a regional basis and are generally proportional to wet sulfate deposition. However, some of the largest
changes in alkalinity are too large to be caused by acid deposition
alone. Using historical data for New York lakes, varying rates of change
in alkalinity could be inferred, depending on the assumptions made:
only one method showed no overall reduction. It was established that
acidification of lakes is accompanied by declining fish populations.
However, the role of acid deposition in the decline of red spruce in
high-elevation forests of the eastern United States could not be
evaluated.[22]

Following a public policy meeting at the Mary Flager Cary Arboretum
in Millbrook, New York, a group of six scientists formed an ad hoc committee to determine the extent of scientific agreement on major aspects
of the acid rain debate. Six governmental reports—three from NAS
(1981–84) and three published in 1984 by EPA, OSTP, and OTA respectively—were analyzed. The following areas of agreement were evident:

1. More than 90 percent of acid deposition is due to man-made
    emissions of SO_2 (two-thirds) and NO_x (one-third). These pol-

lutants are frequently transported hundreds of miles.

2. Acidification has been found where sensitive surface waters receive acid deposition. In aquatic systems species are eliminated. Nutrients are leached from soils, contributing to forest decline. Atmospheric visibility is reduced and materials are corroded.

3. Reducing emissions of SO_2 will reduce deposition of sulfur in a commensurate manner, reducing the effects mentioned above.

4. Adequate scientific information exists to select emission reduction strategies.[23]

The long-awaited interim report of NAPAP was finally released in September 1987, two years behind schedule.[24] NAPAP represents a coordinated effort on the part of the U.S. government to synthesize the scientific evidence on acid rain as the basis for a policy response. As the official report of the administration, its conclusions aroused much curiosity. However, they are so inconsistent with all of the other assessments currently available that a real credibility problem resulted. The report consists of four volumes, of which the first is the executive summary. It appears to be a highly politicized document. The Canadians were particularly outraged by the summary and immediately prepared a document refuting many of the conclusions, which they claim misrepresent the information contained in the main body of the report. In addition to some specific comments, the critique counters five basic conclusions that it sees as totally misleading and inaccurate:

1. The effects of acid rain are neither widespread nor serious.

2. There will be no abrupt changes in the effects of acid rain for the next several decades.

3. Emission levels of SO_2 have been nearly constant since 1920 and will decrease substantially over the next thirty to forty years through the application of new technologies due to market forces.

4. The effects of acid rain are less than were anticipated ten years ago.

5. Sufficient uncertainty remains to preclude a determination of the need for or the nature of abatement strategies.[25]

One of the most serious points of contention has been the use in the NAPAP conclusions of pH 5.0 as the definition of a lake that is affected by acid deposition. By choosing this value as opposed to pH 6.0, at which adverse biological effects can be detected, the number of affected lakes in eastern Canada is reduced from 150,000 to 14,000. This and other scientific considerations have precluded endorsement by reputable scientific organizations. There is good reason to believe that this report is not an entirely objective representation of the current state of knowledge on the causes of acid deposition. The conclusions appear to contradict a 1981 NAPAP report (summarized in the appendixes of this book), which anticipated a slight increase in annual emissions of SO_2 and a 16 percent increase in NO_x (see table A-12). This earlier report also acknowledged that many lakes are experiencing acid-induced stress, which may result in fish kills (see tables B-1 and B-7). It may be significant that the U.S. Office of Science and Technology Policy embraces the interim report, whereas in 1984 it reported that acidification of lakes had accelerated in recent decades and was accompanied by the disappearance of aquatic biota. At that time it expressed concern at potential damage to soil microorganisms and plants and recommended that additional steps be taken to control emissions of sulfur and nitrogen oxides.[26]

Conclusions

There is general agreement among the scientific community about the seriousness of the acid rain problem. It is also agreed that, although specific source-receptor relationships have not been established, strong regional relationships between SO_2 emissions and acid deposition exist. Substantial damage to some aquatic ecosystems in North America has been established, although the extent to which damage to forests can be attributed to acid rain is not known. The largest area of uncertainty concerns the quantitative relationship between causes and effects and, thus, what benefits will be realized at different levels of control.

This view of the scientific nature of the acid rain debate is shared by the Canadian government, which has taken action to reduce emissions accordingly (see chapter 6). In contrast, the Reagan administration still maintains that there is too much uncertainty to justify additional controls, and it uses selective scientific data to support this position.

Chapter Three
Negotiations On Acid Rain

⁗⁗

Introduction

The current disagreement between the United States and Canada over acid rain illustrates the difficulties involved in resolving a bilateral dispute between two nations. Canada and the United States are closely allied on many issues of mutual concern and have cooperated in the past to resolve conflicts. However, the acid rain issue presents unusual complexities and, based upon the current status of the formal negotiations, it is unlikely that an agreement will emerge in the near future. This chapter traces the origins, progress, and present status (spring 1988) of the acid rain negotiations between the United States and Canada and examines the major areas of disagreement.

The Beginning of Negotiations

The foundation for agreements on environmental matters between Canada and the United States was laid on January 11, 1909, with the signing of the Boundary Waters Treaty. This agreement established the International Joint Commission (IJC) and adopted a number of fundamental principles in the regulation of international waters separating the two countries.[1] The accomplishments of the IJC and its role in alerting the governments to the threat of acid rain are described in chapter 8.

Momentum toward a bilateral agreement on acid rain did not develop in the United States until 1978, when the Great Lakes Water

---

This chapter was written by Robert B. Stewart.

Quality Agreement was concluded under the auspices of the IJC. One part of the agreement stipulated that each government would take appropriate measures to control, abate, and prevent pollution from airborne contaminants, and consult on remedial measures. This section had been included in the agreement as a result of IJC studies during 1972–78 showing that a high proportion of pollutants entering the Great Lakes came from the atmosphere. Congress also placed a rider on the Foreign Relations Authorization Act of 1978, directing the Department of State to begin negotiations with Canada on an air quality agreement.[2] This congressional concern was prompted both by the IJC reports and by Canadian plans to construct two new oil-fired thermal generating plants across the border from Montana and Minnesota.[3] In October 1978 the Bilateral Research Consultation Group on the Long-Range Transport of Airborne Pollutants was formed subsequent to correspondence between the Department of State and the Canadian embassy. This group was given the task of consulting on ongoing research efforts and facilitating an exchange of technical information on the long-range transport of air pollutants.[4]

Following a further exchange of notes between the governments in November 1978, informal bilateral talks took place on December 15, 1978, and again on June 20, 1979. At these talks discussion papers on the major aspects of the issue were exchanged. The discussions led to a joint statement released in Ottawa on July 26, 1979,[5] in which the parties made reference to two earlier environmental agreements. One, the Boundary Waters Treaty, obligates both governments to ensure that "boundary waters and waters flowing across the boundary shall not be polluted on either side to the injury of health or property."[6] Second, both governments supported Principle 21 of the 1972 Stockholm Declaration on the Human Environment, which declares that states have the "responsibility to ensure that activities within their jurisdiction or control do not cause damage to the environment of other states or areas beyond the limits of national jurisdiction."[7] The joint statement commits both governments to the development of a cooperative bilateral agreement on air quality, moving the discussions beyond the informal stage. This commitment is as yet unfulfilled, but remains binding on both partners.

The statement concludes with a list of principles or practices that were to be addressed in the formal negotiations. These include prevention and reduction of harmful transboundary air pollution; development and implementation of control strategies; expanded notification and consultation practices; increased technical research and exchanges of scientific information; expanded monitoring and evaluation; long-term environmental assessments; consideration, as part of an agreement, of institutional arrangements, equal access, nondiscrimination, liability, and compensation; and actual implementation measures.

Canadian hopes for fast progress in reaching an agreement were diminished when, in February 1980, the Carter administration announced a program to convert over one hundred oil-fired power plants, located in different parts of the country, to coal. Once implemented, the plan would have substantially increased U.S. sulfur dioxide emissions. Although President Carter realized that his decision would harm U.S.-Canadian environmental relations, the goal of energy self-sufficiency took precedence. The conversion plan ultimately failed in Congress, but damage to the negotiations had been done.[8]

The Memorandum of Intent on Transboundary Air Pollution

Despite the setback, negotiations continued and a Memorandum of Intent on Transboundary Air Pollution (MOI) was signed on August 5, 1980. This agreement committed both governments to develop a bilateral agreement and, in the interim, to take preliminary actions to control transboundary air pollution. The major provisions are: "(1) to develop a bilateral agreement which will reflect and further the development of effective domestic control programs and other measures to combat transboundary air pollution; (2) to facilitate the conclusion of such an agreement as soon as possible; and (3) pending conclusion of such an agreement, to take interim actions available under current authority to combat transboundary air pollution."[9]

The MOI established a joint coordinating committee to begin preparatory discussions immediately and formal negotiations no later than June 1, 1981. The committee would oversee work groups in charge of investigating particular issues. Generally, the MOI formalized the 1979 joint statement. It went further in the control measures section, however, since it committed both governments to reduce transboundary pollution by developing national control measures and enforcing existing controls. The exact language of these provisions is important to note. Both countries agreed to: "(a) develop domestic area pollution control policies and strategies, and, as necessary and appropriate, seek legislative or other support to give effect to them; (b) promote vigorous enforcement of existing laws and regulations as they require limitations of emissions from new, substantially modified and existing facilities in a way which is responsive to the problems of transboundary air pollution."[10]

Negotiations during the First Reagan Term

During the spring of 1981 progress slowed. The new Reagan administration appointees in the Interior Department and the Environmental Protection Agency (EPA) disagreed with their Canadian counterparts over the extent and danger of acid rain, asserting that further research was necessary prior to implementing costly control measures. The Canadian minister of the environment, John Roberts, repeatedly expressed concern that the United States was not living up to the provisions of the MOI.[11] Roberts charged that the use of cost-benefit analyses to determine control levels, as suggested by the United States, was unacceptable when applied to transboundary pollution. He further complained that no new interim controls, as specified in the MOI, were being implemented or even contemplated by the U.S. government. Also surfacing was the suggestion that Canada might use its supplies of hydroelectric power, a significant amount of which is exported to the U.S., as a lever to force U.S. acceptance of a bilateral acid rain treaty.[12] In April, however, Roberts met with Secretary of the Interior James Watt and was assured that negotiations would begin on schedule and that no changes would be made to the U.S. Clean Air Act that could harm Canada. (The act was up for reauthorization and had been viewed by the new administration as a prime vehicle for granting regulatory relief to industry.) Secretary Watt promised that the government's power to control acid rain would not be curtailed.[13]

The growing reluctance by the Reagan administration to begin formal negotiations was in part responsible for a hearing on May 20, 1981, before two subcommittees of the House Committee on Foreign Affairs.[14] The U.S. chairman of the Coordinating Committee on Transboundary Air Pollution, Ray C. Ewing (who was also the deputy assistant secretary of state for European affairs), was one of the witnesses. He assured the committees that formal negotiations would begin as scheduled the following month, and that Congress would be consulted fully during the course of these negotiations. This last assurance, according to committee staff, was never fulfilled.

Outside of recounting past actions, the one notable point made by Ewing was that more research was needed to confirm the cause-and-effect relationship in the acid rain issue.[15] This position was later to emerge as the principal stand of the U.S. negotiating team and is a primary reason for the presently stalled negotiations. Mr. Ewing noted a close link between the reauthorization of the Clean Air Act and the negotiation of a bilateral transboundary air pollution agreement,[16] but did not say how the administration would handle the linkage. In response to concerns from committee members that the negotiations could proceed indefinitely and therefore would not be concluded in conjunction with the reauthorization of the Clean Air Act, Ewing could only reaffirm the administration's commitment to conduct serious negotiations on the issue and to endeavor to conclude an agreement as soon as possible. Over the next several months it became clear that the Reagan administration did indeed intend to link the Clean Air Act reauthorization with acid rain negotiations, thus further reducing the chances of a timely bilateral agreement.

The first formal negotiating session took place on June 23, 1981. The meeting, and a second session in November, consisted primarily of an exchange of ideas and a review of the available scientific data. Draft portions of an agreement were exchanged. At the third session in February 1982 Canada submitted its first formal proposal: Canada would reduce sulfur dioxide emissions by 50 percent by 1990 if the United States would take similar action. The United States preferred a longer time frame for developing solutions, with U.S. policy to be shaped in response to future research. The Canadians agreed with the need for a flexible, long-term strategy, provided it would not be used to prevent interim action.[17] Specifically, the Canadians criticized the lack of any proposals for new air pollution control measures from the Reagan administration. The Canadian negotiators felt that their government was taking concrete steps to reduce sulfur dioxide emissions from Canadian industries and utilities, while the United States was attempting to lessen existing environmental controls. The Reagan administration's deregulation program, like the Carter coal plan before it, took priority over acid rain within the U.S. administration. While the Canadians sought a long-term agreement and immediate control measures, the United States proposed continued cooperation in research, which would clarify the uncertainties and lay the groundwork for effective programs to deal with the problem.

In October 1982 Secretary of State George Shultz met with Canadian External Affairs Minister Allan MacEachen concerning the negotiations, and agreed to exchange papers on key points relating to acidic deposition. In January 1983 these papers were formally exchanged. The Canadians reiterated their claim that sulfur dioxide emissions from coal-burning plants were the major causes for fish kills in freshwater lakes, as well as crop and forest damage. Again, a mutual 50 percent reduction in emissions was suggested. The U.S. position remained that more research was required before expensive control measures would be considered.[18]

April 1983 saw another meeting between External Affairs Minister

MacEachen and Secretary Shultz in Washington, but little progress was made toward resolving the dispute. They decided, however, to have scientific advisers compare notes and make plans for another meeting in the summer.

Hopes for a timely resolution of the issue were revived when President Reagan appointed William D. Ruckelshaus to succeed Anne Burford as administrator of the EPA in May 1983. The president directed that acid rain be the first priority for the new EPA administrator. Despite cutbacks in other areas, funding for acid rain research was doubled.[19] Ruckelshaus stated at the time that: "My understanding now is that there is no question that there is a problem of acid deposition that impacts on certain lakes in the northeastern part of this country and in Canada and that a major contribution to the cause of that is man-made."[20] But he also noted that he was uncertain how to go about fashioning a program that might provide relief for acid rain-stricken areas.

Throughout the summer of 1983 Ruckelshaus worked to develop a control program that would have a chance of being backed by a large enough coalition of interests. By September he had developed a plan that called for modest reductions in emissions from coal-fired power plants. When presented to the Cabinet Council on Natural Resources and Environment, the plan met stiff opposition from Budget Director David Stockman and Energy Secretary Donald Hodel. Stockman and Hodel blocked any new controls, stating that Ruckelshaus's proposal would be too costly to electric power companies and their rate-payers.[21] In hearings before the U.S. Senate Committee on Environment and Public Works, Ruckelshaus accepted the president's judgment to defer controls, and pointed out that an eventual SO_2 control strategy would require several elements: improved information on what new controls would buy; a reasonable consensus among the most affected parties; and a system for sharing the cost of controls by the entire nation.[22] The *Washington Post* reported that Ruckelshaus told the Canadians not to expect action soon.[23]

In his State of the Union Address in January 1984, President Reagan restated the administration's "research before action" strategy. In response, the Canadian government sent a formal note registering its disappointment with the U.S. position and asking how the U.S. government planned to honor previous commitments, which had been restated by President Reagan during his visit to Canada in 1981, to control transboundary air pollution. At the time the president had stated before the Canadian House of Commons: "We want to continue to work cooperatively to understand and control the air and water pollution that respects no borders." The Canadian note three years later accused the U.S. government of no longer honoring this commitment: "The continued delay in adopting effective abatement measures is not acceptable to Canada. Canada considers that the decision [to limit action to increased research] fails to take full account of U.S.A. undertakings and ignores protecting the North American environment."[24]

The Special Envoys on Acid Rain

After the election of a more conservative administration in 1984, the Canadian government toned down its attacks on the Reagan administration and is now trying to advance negotiations through quiet diplomacy. On March 17 and 18, 1985, Prime Minister Brian Mulroney and President Ronald Reagan met in Quebec City to discuss a wide range of bilateral issues. At that meeting acid rain was recognized by both leaders as a serious concern affecting bilateral relations. Each agreed to appoint a special envoy to review the issue and report to them prior to their next summit meeting in the spring of 1986. Drew Lewis, former secretary of transportation, and William Davis, former premier of Ontario, were assigned four specific tasks: (1) to pursue consultation on laws and regulations related to pollutants thought to be linked to acid rain; (2) to enhance cooperation in research efforts, including research on clean-fuel technology and smelter controls; (3) to pursue means to increase exchange of relevant scientific information; and (4) to identify efforts to improve the U.S. and Canadian environment.[25]

After a year of studying the problem and meeting with representatives of all the interested parties—government, industry, and nonprofit organizations—they concluded that:

1. *Acid rain is a serious problem in both the United States and Canada*. Acidic emissions transported through the atmosphere undoubtedly are contributing to the acidification of sensitive areas in both countries. The potential for long-term socioeconomic costs is high.

2. *Acid rain is a serious transboundary problem*. Air pollutants emitted by sources in both countries cross their mutual border, thus causing a diplomatic as well as an environmental problem.

3. *At the present time, there are only a limited number of potential avenues for achieving major reductions in acidic air emissions*, and they all carry high socioeconomic costs.[26]

Keeping in mind that the envoys' mandate was not to find a final solution to the bilateral problem of acid rain, recommendations were made in three general areas:

1. *Innovative control technologies.* The U.S. government should implement a five-year, $5 billion control technology commercial demonstration project, funded jointly by the federal government and industry. Projects should be selected according to their ability to produce the greatest emissions reductions at the least cost at the greatest number of facilities. This research effort should be overseen by a panel headed by a senior U.S. cabinet official and should include representatives from the Department of State, state governments, and Canada.

2. *Co-operative activities.* Opportunities should be found in existing clean air legislation in each country for addressing environmental concerns related to transboundary air pollution. Diplomatic channels should be used to convey contemplated changes in these laws or regulations to the other country. Acid rain should remain high on the agenda in meetings between the president and prime minister, and a bilateral advisory group on transboundary air pollution should be established using both diplomatic and environmental management officials.

3. *Research.* The following areas were identified as those that would help to dispel some of the remaining uncertainty associated with acid deposition: development of a dry deposition monitoring network; development of models for predicting watershed responses to acid deposition; quantification of the relationship between changes in surface water chemistry and changes in aquatic biota; investigation of the potential link between forest decline and acid rain; quantification of the extent of current damage to structures; and determination of the potential health effects associated with increased solubility of heavy metals in acidified water.[27]

Because the report acknowledged the seriousness of the transboundary environmental problem but did not recommend immediate action, it has received a mixed reception. Canadians were particularly disappointed. Nevertheless, Canadian Ambassador Allan Gotlieb lobbied hard to secure White House endorsement prior to the March 1986 summit meeting between President Reagan and Prime Minister Mulroney. However, the whole process appears to have been an exercise in diplomatic niceties, without achieving concrete results. At the time, $800 million had been secured for clean-coal technology research in the United States, but Congress has not appropriated any more money since then. In his 1988 budget the president did request the federal government's share of a $5 billion clean-coal technology development program. However, the Canadians responded by saying that the request did not fulfill the commitment made by the president to implement the terms of the special envoys' joint report. They see the initiative as falling short in three major areas: (1) it will not provide any measurable reduction in transboundary pollution prior to 1995; (2) no significant reductions in transboundary air pollution flows will be apparent for at least twenty-five years, whereas clean-coal technology initiatives that met the criteria delineated in the envoys' report could result in cleaner, more cost-effective technologies within ten years; (3) of the $6.8 billion requested, $5.3 billion is earmarked for projects that meet one of the technical criteria, and only $1.7 billion of expenditures meet all of the technical criteria of the envoys' report.[28]

In April 1987 President Reagan said he would consider negotiating an acid rain accord. Responding to this initiative, the following month the Canadians submitted a conceptual paper as the basis for negotiation. On January 25, 1988, following extensive interagency review in the United States, a joint meeting was arranged in Washington between U.S. and Canadian officials. A full draft accord was submitted by the Canadians; but with the U.S. side still refusing to consider reduction targets or a timetable for controls, there is little possibility of a settlement emerging from this meeting.[29]

The Canadian negotiating position remains unchanged. A Canadian diplomat summed it up in three points on which agreement is sought from the United States: (1) a significant reduction in sulfur dioxide emissions; (2) a commitment to a deposition maximum of eighteen pounds of sulfate per acre per year; and (3) the creation of a bilateral mechanism to review new scientific information and to monitor the state of the environment.[30]

In the first edition of this book we stated that "the impasse within the Reagan administration over national as well as international acid rain controls persists to this day (spring 1985). Congress also remains unable to achieve a consensus on acid rain legislation. No action was taken during the election year, and the 1984 election did not push the issue to the forefront of public attention, thus precluding a national debate of the issue and a possible settlement in the near future." The situation in spring 1988 has changed very little. As discussed in chapter 7, Congress is somewhat closer to coming to a bipartisan agreement over acid rain controls, but the administration persists in its position that more research is needed. In fact, as discussed in chapter 2, the official position of the administration, as delineated in the interim report of the National Acid Precipitation Assessment Program, appears to be a step backward. It seems unlikely that the Reagan administration will change its position during the last year of its tenure.

Issues and Perspectives

Scientific Uncertainties

The present impasse in the negotiations is attributable to disagreements over a wide range of issues. As chapter 2 shows, scientific experts generally agree on what is known about acid rain and what can and cannot be concluded from the available information. The major differences between the U.S. and the Canadian positions arise from different views about the urgency of corrective action and whether damage can be reversed. The United States has adopted the stance that gaps in knowledge about the chemistry and transport of acid deposition make it unwise, at this time, to impose costly legislative remedies to control sulfur dioxide emissions. Limited resources should be spent on research and development of more effective control equipment rather than on a control program of unproven effectiveness.

There has been some evolution in the U.S. position. In the early years of the Reagan administration, highly placed officials such as former EPA administrator Anne Burford pointed out that "several rigorous studies cast doubts on the theories" that the major causes of acid rain are coal-fired industries.[31] Her successor at EPA accepted the consensus of scientific assessments that man-made emissions are a major cause of acid rain. He added: "If acid rain controls were cheap, there wouldn't be any disagreement on the science."[32] Even so, he was unable to convince the president and his advisers that additional air pollution abatement measures should be taken to reduce the long-range air pollution. William Ruckelshaus was replaced as EPA administrator by Lee Thomas, whose views are much more in tune with the administration's position. Thomas has testified before Congress that the Clean Air Act in its present form provides adequate protection against air pollution. He opposes all the major provisions for strengthening the act, maintaining that additional controls for acid rain precursors are not necessary at the present time.

The Canadian position is that interim controls are needed to protect the environment from irreversible damage while further studies are conducted. Research is needed to evaluate and improve abatement efforts, but not as a substitute for control measures. Furthermore, there is a perception among the Canadians that some U.S. environmental officials believe only those studies that support the U.S. position. Canadians point out that U.S. officials discounted a major EPA study on acid rain that strongly linked emissions from midwestern coal-burning utilities to the destruction of lakes and their fish populations in both the United States and Canada.[33]

Before a meeting of the U.S. Air Pollution Control Association, Canadian Environment Minister Roberts declared: "In Canada we are deeply disappointed with the state of negotiations. . . . The foot-dragging and interference in the development of scientific information has reached frustrating proportions. The administration's rejection of our proposal to reduce sulfur dioxide emissions in eastern North America by 50 percent by 1990 and a clear indication that it may be some considerable period of time before it will even be able to discuss control actions, is a bitter pill to swallow."[34]

Since then, even stronger accusations have come from Canadian officials. In a speech before the National Academy of Sciences in Washington, D.C., Raymond Robinson, head of Canada's Federal Environmental Assessment and Review Office, accused the Reagan administration of "blatant efforts to manipulate acid rain work groups."[35] Robinson also charged that the Reagan administration was suppressing scientific information concerning acid rain, that money for mutually agreed-upon clean-up programs was being withheld, and that the United States was still not cooperating in the spirit of previous agreements and had not committed itself to the intent of the MOI to conduct serious negotiations.

In a 1984 demarche to the U.S. government, Canada pointed out that over three thousand scientific studies on acid rain have been conducted, resulting in "sufficient scientific evidence . . . by prestigious scientific bodies in North America and Europe on which to initiate controls programs." The press release summarizing the official note adds that the available evidence has been found sufficient to justify controls in many countries: "Like Canada, Germany, Finland, Sweden, Norway, Denmark, France, Austria, and Switzerland have all agreed, on the basis of the available and overwhelming scientific evidence, to adopt programs to cut back sulphur dioxide emissions."[36]

The recent release of the interim report of the Interagency Task Force on Acid Rain appears to be a step backward on the part of the Reagan administration.[37] In contrast to previous scientific assessments, and inconsistent with the spirit of the special envoys' report, it concludes that acid rain is not a serious problem and that it is too early to determine whether abatement action is necessary. In a letter to EPA administrator Lee Thomas, Canadian Minister of the Environment Tom McMillan characterized the report as "flawed, incomplete and misleading."[38]

Format of the Agreement

A bilateral acid rain agreement could be in the form of a formal treaty or an informal executive agreement. The Canadian government prefers a formal treaty for several reasons. A treaty requires the consent of the U.S. Senate and becomes national law. Under these circumstances U.S. clean air laws would be required to reflect the provisions of the treaty (which the Canadians hope will contain strict source emissions standards) and would be enforceable in U.S. courts. Similarly, a treaty containing such standards would provide for clear, definable avenues whereby Canadians or Americans could contest particular pollution control actions or inactions. A treaty would also facilitate the Canadian federal government's control over provincial actions (subject naturally to political considerations). At present, control of ambient air quality and stationary source emissions are provincial responsibilities. As discussed in detail in chapter 6, control authority only devolves upon the Canadian federal government when there is an international obligation or a health danger. An informal executive agreement would not give the Canadian federal government such authority.

The treaty format, however, has its drawbacks. In the United States the treaty approval process can be long and drawn out. The Senate has been reluctant in recent years to approve formal treaties. Given past failures of treaties concerning fishing rights, maritime boundaries, and the Law of the Sea, an acid rain treaty may never receive Senate approval. A treaty is also less flexible than an agreement. Incorporating specific emissions standards in a treaty would inhibit both governments' flexibility in shaping future environmental regulations. A requirement for scrubber technology on plants would discourage development of new and better control technologies. Similarly, a treaty provision setting emissions levels would be hard to change if new information became available suggesting different levels.

On the U.S. side, an executive agreement can be acted upon by the president alone. The president could enter into an agreement in order to diffuse charges of inaction and improve bilateral relations without waiting for legislation introducing a new air pollution control policy.

It is unlikely that the current administration will opt for this approach. But even if a new administration were politically more inclined to reach an agreement with Canada, it would have to examine carefully whether meaningful reductions in acid deposition can be reached within the framework of existing legislation. This point is addressed later in this chapter in the analysis of Section 115 of the Clean Air Act.

In either case there would still remain the question of enforcement. Canadians point to the recent unwillingness of the EPA to enforce existing pollution standards. Conversely, the U.S. government has criticized Canada for not using available emission control technology. (Since that time Canada has made significant progress in strengthening its clean air laws; see chapter 6.) A solution would be to establish an independent advisory commission to oversee domestic actions in both countries and report its findings of compliance or noncompliance. Though neither country would be inclined to grant such a commission enforcement powers, an independent body would be valuable as an impartial way for evaluating compliance. Another alternative would be to make each country's environmental control agency responsible for oversight. The agencies could make reports to their respective legislative bodies and chief executives as to compliance with the agreement. Whatever alternative is selected, it will work only if domestic law backs up the international agreement.

Political Considerations

Political Developments The elections of 1980 had an important effect on the progress of the negotiations. The Canadian government became concerned that the Reagan administration would not faithfully carry out agreements reached with the Carter administration. Although President Reagan assured former prime minister Trudeau that the United States is committed to achieving a transboundary air pollution agreement, other statements and actions of the president and his associates indicate that the Reagan deregulation program has a higher priority. The Reagan administration has argued for relaxing provisions in the Clean Air Act; has cut back appropriations and staffing for the EPA; has refused to consider various international agreements on fishing rights and the Law of the Sea; and has repeatedly called for continued research prior to implementing any control measures to reduce sulfur dioxide emissions. These administration attitudes and actions have strained harmonious bilateral relations. The 1984 election of Brian Mulroney as prime minister of Canada and the return of Reagan to the White House may have led to better personal relations between the heads of state, but it has not produced significant progress on the acid rain issue.

Environmental Perspectives   Polls have shown that Americans and Canadians alike are concerned about environmental protection. Concern tends to be highest when a problem is viewed as affecting the entire nation and posing a direct threat, in particular to human health, to large numbers of individuals. The evidence on acid rain in either country does not suggest that these conditions are met. But perceptions are a different matter. Canadians definitely view acid rain as a serious national threat. A high degree of public awareness has helped in building a national consensus and surprisingly strong cooperation between industry and government in dealing with the problem. Many Canadians live downwind from U.S. air pollution and therefore are pressing hard for a bilateral air quality agreement. Concern in the United States is likely to increase when coal-fired plants are in full operation at Poplar River, Saskatchewan, and Atikokan, Ontario (both within fifty miles of the U.S. border), bringing transboundary pollution to areas in Montana and Minnesota. Although some Canadians still view their wilderness area as unlimited, concern about its future will increase further should Canadian forests (supplying pulp and paper mills) be damaged and harm to lakes (for fishing and tourism) become more widespread.

American environmentalists are equally concerned over environmental degradation, but attention to acid rain is not yet as intense as the earlier response to such dangers as the use of pesticides and industrial chemicals, water pollution, local air pollution, habitat destruction, and species preservation. During the 1970s Canada followed the U.S. lead in environmental policy. For years the acid rain problem depended on Scandinavian leadership. Acid rain is now an issue around which a variety of Canadian groups, including environmentalists, have coalesced, and Canada has taken the lead in North America. In the long run, however, U.S.-Canadian environmental relations must be built around the fact that both countries share the same atmosphere and environment.

Perspectives on Bilateral Relations   Bilateral relations take on different levels of importance in Canada and the United States. For Canada, relations with the United States are the most crucial aspect of foreign policy. Most Canadians live near the U.S. border and are inundated by U.S. culture, economics, and politics through television, movies, magazines, and newspapers. The United States is Canada's principal trading partner, and the two nations are closely allied on defense matters. Virtually everyone in Canada is aware of the current state and scope of U.S.-Canadian bilateral relations.

In comparison, most Americans have at best only a limited knowledge of Canada and U.S. relations with Canada. The vast range of U.S. international commitments relegates Canada to a less-than-premier position. Bilateral relations with Canada are more often than not considered to be of secondary or lesser importance in U.S. foreign policy. Often U.S.-Canadian cultural, economic, and political similarities, coupled with Canada's relatively small population, have produced a condescending attitude on the part of U.S. foreign policymakers. This has led to the erroneous assumption that Canada will always support U.S. foreign policy initiatives in order to maintain harmonious bilateral relations. The Vietnam War showed that this is not always the case, and continued persistence in this assumption will be detrimental to the progress of the acid rain negotiations. Past bilateral agreements have addressed either mutual concerns (e.g., boundary waters) or issues that were of predominant interest to the United States (e.g., certain fishing and maritime boundary treaties). In the case of acid rain, the United States may have to shoulder more of the costs of control and the Canadians may reap more of the benefits. This will test the nature of cooperation between the neighboring nations.

Linkage Presently, as a matter of principle, neither country favors linking acid rain to other bilateral issues. Both believe that bilateral disputes should be resolved on their individual merits and not by the threat of action (or inaction) by one of the parties in regard to nonrelated concerns. Even so, if negotiations remain stalled Canada may feel compelled to seek redress in some indirect way for the costs believed to be inflicted on the Canadian environment by American industries and utilities. However, both sides can employ this method, and by doing so could damage bilateral relations on the entire range of issues between the two countries.

Although linkage is undesirable, it cannot be overlooked. Neither nation can realistically be expected to give ground on one bilateral issue without receiving satisfaction on other issues. If Canadian-U.S. relations are to remain cooperative and friendly, both sides must realize that the costs of not concluding an acid rain agreement extend far beyond that one issue. Canada and the United States each have an interest in a variety of actions under consideration by the other government. The United States, for example, is interested in Canada relaxing the restrictions contained in its National Energy Policy and the Foreign Investment Review Agency; in continuing close mutual defense efforts; and in obtaining extended rights to Canadian offshore fishing areas. Canada, on the other hand, is concerned with transboundary air pollution; Great Lakes water quality; the Garrison diversion project; and plans to build a major oil refinery and supertanker port at Eastport, Maine, as well as other environmental and trade issues. Both nations, in fact, view trade and associated economic issues as areas of mutual concern.

Canadian Lobbying Efforts   The lack of progress on the diplomatic front in concluding a bilateral acid rain agreement has led the Canadians to pursue other avenues by which to influence U.S. policy on this matter. How effective lobbying will be is as yet unclear. Canadian Environment Minister Roberts has traveled extensively in the United States, speaking to the press and various professional organizations and publicizing Canadian concerns for continued, constructive negotiations. The Canadian Environmental Department also sponsored two controversial films on acid rain that were distributed to interested organizations within the United States. Even though the Canadian Department of External Affairs has discouraged direct participation of cabinet members and civil servants in the American political arena, various officials have testified before congressional committees on reauthorization of the Clean Air Act. A prominent legal firm in Washington has been engaged by the Canadian embassy to monitor environmental issues in Congress that relate to Canadian concerns and to advise the embassy on appropriate lobbying action.[39] Acid rain awareness programs have been started at fifteen Canadian consulates in the United States, and tours have been arranged for legislators, staff members, and journalists to visit acid rain-damaged areas in Canada.

Perhaps the most interesting tactic has been the formation of the Canadian Coalition on Acid Rain (CCAR), the first registered Canadian lobbyist in the United States working for a nongovernment, nonbusiness, citizens' organization. Representing more than fifty separate organizations with membership approaching two million Canadians, CCAR has furthered efforts to publicize the acid rain issue in the United States and pressure American legislators to adopt a regulatory program on acid rain as part of the reauthorization of the Clean Air Act.[40] So far these lobbying efforts have had little visible effect upon the course of the formal negotiations. It is perhaps due to a growing sense of desperation that the Canadians have adopted this type of strategy. Reasoning that the formal negotiations are stalled, Canada is trying out new methods to influence U.S. policy on acid rain.

Section 115 of the Clean Air Act

An important issue in diplomatic negotiations has been the question of the usefulness of existing provisions, in particular Section 115 of the Clean Air Act, as a means for controlling acid rain between Canada and the United States.[41] The Act states:

> a. Whenever the Administrator, upon receipt of reports, surveys, or studies from any duly constituted international agency, has reason to believe that any air pollutant or pollutants emitted in the United States cause or contribute to air pollution which may reasonably be anticipated to endanger public health or welfare in a foreign country, or whenever the Secretary of State requests him to do so with respect to such pollution which the Secretary of State alleges is of such a nature, the Administrator shall give formal notification thereof to the Governor of the State in which such emissions originate.
>
> b. The notice of the Administrator shall be deemed to be a finding under Section 110(a)(2)(H)(ii) which requires a plan revision with respect to so much of the applicable implementation plan as is inadequate to prevent or eliminate the endangerment referred to in subsection (a). Any foreign country so affected by such emissions of pollutants shall be invited to appear at any public hearing associated with any revision of the appropriate portion of the applicable implementation plan.
>
> c. This section shall apply only to a foreign country which the Administrator determines has given the United States essentially the same rights with respect to the prevention or control of air pollution occurring in that country as is given that country by this section.[42]

The reciprocity provision contained in subsection (c) has led to Canadian action. In December 1980 the Canadian Parliament passed Bill C-51 (an amendment to the Canadian Clean Air Act), which is remarkably similar to Section 115 of the Clean Air Act. Upon its passage the EPA administrator acknowledged that it met the reciprocity requirement, thus clearing the way for the potential use of Section 115.

The Canadian federal government, which has limited constitutional powers in environmental policy, had previously used its international obligations to involve itself in provincial air pollution cases. Actions taken include reductions of sulfur dioxide at INCO, Ltd., monitoring of discharges at the Poplar River power plant, and scaling down the size of the Atikokan power plant. This cooperation between the federal and provincial governments was intended to show the United States that Canada has taken action to reduce its own sulfur dioxide emissions. Although the federal government is not in a position to force a province to implement pollution control measures with which it does not agree, continuing cooperative efforts to reduce emissions by 50 percent suggest that this does not create a barrier to further progress.

Another question is whether or not Section 115 and Bill C-51 are the appropriate means for dealing with transboundary air pollution. As an interim action their use would be appropriate. The MOI of 1980 specifically required that each country take interim actions to control transboundary air pollution, utilizing existing domestic legislation. Thus, the use of these provisions would be an appropriate means to meet an international obligation the United States entered into in 1980, until such time as a formal agreement has been concluded.

But there are questions concerning the use of Section 115. For example, what is meant by the words "endangers public health or welfare in a foreign country"? As long as scientific uncertainties remain or as long as the administration does not recognize a direct cause-and-effect relationship between sulfur dioxide emissions in the United States and acid rain in Canada, it is unlikely that Section 115 will be utilized. To prevent reevaluation of state implementation plans, the EPA administrator or the secretary of state need only state that no definable problem exists. Under present conditions, therefore, Section 115 of the Clean Air Act will not be used in place of a negotiated, bilateral, transboundary air pollution agreement. If attitudes or administrations change, the EPA, working in concert with affected states (and Canada), could draw up specific measures to implement the provisions of Section 115.

Conclusions

Negotiations, both informal and formal, toward concluding a transboundary air quality agreement have been in progress since 1978. Initially, progress was made in identifying issues, organizing research groups, and concluding interim agreements. Serious setbacks occurred, however, when the United States decided to embark upon power plant conversions to coal and when the Reagan administration began its deregulation drive, under which acid rain became a secondary issue. The negotiations came to a virtual standstill when U.S. negotiators claimed that scientific uncertainties concerning acidic deposition had to be resolved before costly sulfur dioxide control measures could be adopted.

In order to restart productive negotiations, both sides must recognize that the dispute affects the entire range of U.S.-Canadian relations. Leaving this issue unresolved will eventually lead to the deterioration of bilateral relations in other areas. This could result in a de facto linkage to other bilateral issues and the inability to resolve them successfully. A formal treaty or executive agreement on acid rain control will be needed to specify the rights and responsibilities of both sides, giving both nations avenues to address grievances in the other country. Political goodwill and cooperation between Canada and the United States will both be needed to resolve this serious bilateral dispute.

Chapter Four
U.S.-Canadian Research Groups

|||||||||

Introduction

This chapter reviews the history and conclusions of the U.S.-Canadian
research groups that were established when bilateral negotiations
commenced. Their mission was to assess the scientific evidence concerning long-range transport of air pollutants and the effects of acidic
deposition. The bilateral research groups were to provide a shared
scientific base as a starting and reference point for the diplomatic
negotiations on acid precipitation. Over time the process became
highly politicized, and separate conclusions were reached by each
country. Currently there is no formal bilateral research effort with a
specific mandate. Instead, regular meetings between various governmental entities provide for the exchange of information and coordinated project planning.

The Bilateral Research Consultation Group

Initial Efforts: 1978–80

In 1978 the two governments established the United States–Canada
Bilateral Research Consultation Group on the Long-Range Transport
of Air Pollutants (BRCG) to coordinate the exchange of scientific
information and assess current research. The group consisted of ten
U.S. scientists, primarily from the Environmental Protection Agency
(EPA), and ten Canadian members, primarily from Environment Can-

---

This chapter was written by Katherine Wilshusen.

ada. Acid rain quickly became the focus of the group's work.

The group met twice, on July 26 and 27, 1978, at Research Triangle Park, North Carolina, and on March 6 and 7, 1979, at Downsview, Ontario. Two formal reports were produced: *The Long-Range Transport of Air Pollutants in North America*[1] and the *Second Report on the Long-Range Transport of Air Pollutants.*[2] At the time of publication these reports represented the most comprehensive and current assessments of acid rain in North America and were important sources of information for policymakers.

The first report was an introduction to the problem of long-range transport of air pollutants, and it identified acid precipitation as the most significant environmental problem affecting both countries. The report concluded that the net flux of sulfur compounds was from south to north across the border, but most of the total sulfur deposition within each country was probably due to domestic emissions. Cross-border pollution was more significant in Canada than in the United States. Damage to aquatic ecosystems as a result of decreased pH levels had been well documented, but effects upon terrestrial ecosystems were more difficult to quantify. More research was recommended, with emphasis upon causes and effects of acid deposition and identification of effective methods for slowing the rate of future damage. But the report also recommended that interim action should be taken without delay because "it should be recognized that delays in undertaking action to address the presently deteriorating situation will result in greater damage and possible economic costs in the future."[3]

The second report, released in November 1980, updated and quantified much of the information presented earlier. A large section dealt with the use of modeling techniques for establishing patterns of acid precipitation. Models exploring sulfate emissions and deposition were emphasized. The group concluded that realistic models to predict long-range transport and deposition of nitrates were not yet available and were at least a year away. Models suitable for use in control strategies were probably even further away.[4] Better understanding of source-receptor relationships, based on mathematical models, was a key requirement for developing control strategies. The report stated that considerable progress was being made in the field of modeling. But most of the effort was on predicting the behavior of sulfur compounds in the atmosphere, and little work was under way on NO_x and hydrogen ion deposition. The group recommended establishing a set of guidelines for evaluation and comparison of models. A series of work groups on modeling was also recommended. The work of the

BRCG provided the information base for early diplomatic initiatives, in particular the 1979 Joint Statement on Transboundary Air Quality, and the 1980 Memorandum of Intent on Transboundary Air Pollution.[5]

The Memorandum of Intent on Transboundary Air Pollution

The Memorandum of Intent, signed in 1980, included arrangements for developing a mutually acceptable information base on the scientific and technical aspects of acid precipitation. Five bilateral work groups were established to review the scientific evidence and present it in a format that would be useful in negotiations. The groups were to compile a "snapshot" of knowledge to date, assemble and analyze available information on the nature and causes of acid rain, and identify control options. The work groups were composed mostly of government experts, working at the federal level in the case of the United States, and at federal and provincial levels in the case of Canada. Each work group was co-chaired by a representative from each country.

Work Group 1 (Impact Assessment) focused on the effects of acid precipitation on lakes, fish, forests, wildlife, crops, buildings, and human health. The group was to provide information on current and projected impacts in sensitive receptor areas. Work Group 2 (Atmospheric Sciences and Analysis) focused on atmospheric modeling, attempting to trace or predict movement and transformation of pollutants through the air, specifically between source regions and sensitive receptor areas. Work Group 3A was to develop control scenarios based on information from the other groups and also to coordinate the activities of the other groups. Work Group 3B (Emissions, Costs, and Engineering Assessment) was to identify, assess, and price control technologies and quantify pollutant emissions. Work Group 4 was to develop legal documents and institutional mechanisms for the control of acid rain.

The work was divided into three phases. Phase I reports were released in February 1981. The change of administrations in the United States made preparation of Phases II and III of the work group reports difficult, since the new administration was adamant about restricting the role of the work groups. Many members of the work groups were replaced, and conclusions from the Phase I reports, even though they had been agreed upon by both sides, were no longer accepted as a consensus opinion.

James McAvoy, formerly the regional EPA director in Ohio, played the central role in the development and implementation of the Reagan

Table 4-1.   Separate Conclusions of Work Group 1
Concerning Aquatic Ecosystems

Canadian View	U.S. View
Sulfuric acid is the dominant compound contributing to water acidification. Nitric acid contributes most heavily in cases of snowmelt and less so in the eastern part of North America.	Sulfuric acid appears to contribute more than nitric acid.
Studies of lakes indicate higher-than-expected sulfate levels.	No explicit parallel conclusion.
Increased content of sulfate in surface water can be linked to high deposition patterns.	In Quebec, sulfate concentrations in surface waters decrease in parallel with the deposition pattern of sulfates.
Statistically significant decline in pH of aquatic ecosystems since the early 1900s.	Historically, there has been a pH decline since the early 1900s in eastern lake studies.
Present empirical evidence covers a broad spectrum of physical and climatological conditions across northeastern North America and provides a reasonable basis on which to make judgments on potential loading relationships.	Based on this status of scientific knowledge, the U.S. work group concludes that it is not possible to derive quantitative loading-effects relationships.
Reductions of sulfur depositions in precipitation to less than 20kg/ha/yr will protect all but most sensitive areas.	No conclusion.

Historical data indicate population extinction of fish in Adirondacks lakes:
    1930s: 8% pH < 5.0; 10% no fish
    1970s: 48% pH < 5.0; 52% no fish
	The factors causing the population extinction of some fish in the Adirondacks have not been demonstrated.

No explicit parallel conclusion.	In most reported cases, clear relationships have not been established between acidic deposition and observed effects.

Table 4-2.   Joint Conclusions of Work Group 1

Terrestrial Ecosystem Impacts

Acidic and ozone deposition occur at concentrations above background levels at long distances from emission sources. Sulfur dioxide is more of a concern to vegetation in proximity to point source.

Damage to crops and forests as a result of long-range transport of ozone has been documented. Damage to crops has been quantified.

Attention so far is focused on the sensitivity of soils and bedrock because results from studies that address vegetation and ecosystem effects are limited and not well understood.

Human Health and Visibility

Available information gives little cause for concern over direct health effects from acidic deposition. Emphasis in research should be upon indirect effects of metallic contaminants of water and food sources and health implications of recreational activities in impacted waters.

Health hazards and visibility should be viewed as a concern only where ambient air quality standards are violated.

Man-made Structures

Certain airborne chemicals can accelerate deterioration of materials.

It is reasonable to assume that acidic deposition due to long-range transport of air pollutants contributes somewhat to material effects. Current understanding of material decay processes leads to the tentative conclusion that local sources of corrosive pollution mask the effects resulting from long-range transport of acid deposition.

Remedial Measures

Liming is an obvious action to be considered.

Liming may be an effective, temporary means of mitigating the effects of acid deposition upon aquatic ecosystems. Long-term viability and impacts on fish populations need further study.

Terrestrial liming has serious limitations, not the least of which is cost.

Liming may effectively be used in the treatment of municipal water supplies.

administration policy on acid rain. By the end of 1981 most of the new positions had been worked out. But for at least six months the work groups were in turmoil. The Canadians began to complain that no one was showing up at the meetings, and when they did come, they were new faces. Work Group 3A's role was limited to coordinating the activities of the other groups, thus dropping from the agenda the development of control scenarios, which was deemed an inappropriate task for a bilateral group. There were major turnovers in the U.S. membership of Work Groups 1 and 2, and Work Group 1 had three different U.S. co-chairmen in rapid succession. Most of the difficulties were due to the change in administration, but other factors were at work as well. The scientific reviews of the Phase I report of Work Group 1, for example, were quite negative. This raises questions about the quality of the work of the experts involved in this stage of the assessment.

Work progressed slowly. Work Groups 1 and 2 completed their Phase II reports in October and July 1981, respectively.[6] Phase III reports were completed considerably behind schedule, due to problems with preparing interpretations and conclusions. Work Group 1 (Impact Assessment) submitted its final report in January 1983. It reached agreement on most of the information and conclusions, but could not reconcile significant differences of opinion in regard to aquatic impacts. Consequently, separate conclusions were prepared by the Canadian and the U.S. members. Table 4-1 summarizes the points of disagreement between the delegations concerning aquatic impacts. Table 4-2 highlights their joint conclusions in the remaining issue areas.[7]

One of the most significant differences between the Phase II and Phase III reports of Work Group 1 involved the maximum loading limit of pollutants in the atmosphere. The Phase II report had contained a consensus target figure for loading. But the U.S. members working on the Phase III report refused to agree to a target-loading value. The Canadians considered adopting the Phase II value in their final report; in the end they decided to issue a separate conclusion in the Phase III report instead.

Work Group 2 (Atmospheric Sciences and Analysis) completed its Phase III report in November 1982. The members of the group evaluated eight regional-scale atmospheric models of air movement over the North American continent, and also assessed the appropriateness of the methods and assumptions used in these regional models to quantify source-receptor relationships on an annual basis. The major conclusions and recommendations from the report of Work Group 2 are summarized in Table 4-3.

The difference between the reports of Work Groups 1 and 2 is striking. In the case of Work Group 1, sensitive issues concerning aquatic impact could only be reported in separate U.S. and Canadian statements, and these closely echoed each government's official position. Work Group 2, on the other hand, arrived at strong joint conclusions linking areas of emissions to receptor areas. A former member of

Table 4-3. Joint Conclusions on Regional Atmospheric Modeling

Monitoring data in North America show a strong geographical correspondence between a large region of precipitation having low pH and the region of the most intense emissions of sulfur and nitrogen oxides. The region with low pH also corresponds to the region having the highest concentrations and depositions of sulfates and nitrates in precipitation, and in both cases the highest levels are over and immediately downwind of the major source regions.

There are possible natural sources of sulfur and other compounds contributing to the acidity of a region.

Back air trajectory analysis (classifying acid precipitation data by which direction the air mass appeared to be taking) indicates higher sulfate and nitrate concentrations in the air and precipitation when the air mass has passed over areas of higher emissions. This type of analysis cannot distinguish between near and more distant sources, however.

Deposition values at more remote stations demonstrate the existence of long-range transport of air pollutants.

Thirteen proposals have been developed regarding future joint U.S.-Canadian research in the areas of atmospheric modeling, monitoring, and evaluation of results.

Work Group 2 who was not involved in the work on the Phase III report remarked that he was surprised that such definitive conclusions had been reached.[8] The work group was unable to complete a proposed evaluation of current modeling techniques that would identify the most promising one for further development. The group concluded that all models were inadequate and that improvement required, among other things, greatly improved knowledge of actual air movements.

Work Group 3A has not yet submitted a final report and most likely will never do so. This group, composed of the co-chairs of all the work groups, is drawing up proposals about what to do with the information gathered and how it might be incorporated into the negotiating process.

Work Group 3B (Emissions, Costs, and Engineering Assessment) completed its final report in June 1982.[9] The members were to conduct a review of currently available technologies—and their costs—for reduction of SO_2 and NO_x emissions for both new and retrofit installations. This was the most technical of the Phase III reports. Some of the more important conclusions are:

On a per-unit-reduction basis, SO_2 controls may be far more costly for some sectors of industry than for others;

the cost for control equipment is a function of the degree of control desired, and is greater for retrofit installations than for new
installations;

available emission projections predict that between 1980 and 2000
SO_2 emissions will increase 7.6 percent and NO_x emissions will
increase 26 percent;

research and development for SO_2 and NO_x control for combustion sources is currently centered in three principal areas of
activity: improvement in flue gas desulfurization technology, combustion modifications, and fluid-bed combustions. The group agrees
that this research is necessary and recommends that it be supported to the maximum practicable extent.

Peer Review

The peer review of the work group reports presented unusual difficulties. Canada suggested joint review by the Royal Scientific Society
of Canada and the U.S. National Academy of Sciences. The Reagan
administration preferred a national review and concluded that the
White House Office of Science and Technology Policy (OSTP) was better qualified to integrate scientific findings with policy recommendations. George Keyworth, the president's science adviser, was designated to assemble the American review committee.[10] The Canadians
had to settle for their own national review and designated the Royal
Scientific Society of Canada for the task.

Although not provided for in the MOI, it had been expected that
review of work group reports would be conducted jointly to avoid
"competing science" assessments.[11] Even if the OSTP panel were
scientifically objective, the impression of political bias was created,
reducing the credibility of the review. In addition, the decision cast
doubt on the sincerity of U.S. participation in the effort to develop a
bilateral scientific consensus regarding acid precipitation in North
America. Kenneth Hare, the Canadian co-chairman of the U.S.-Canada
Joint Scientific Committee, points out that the Canadian review panel
was composed of U.S. and Canadian members as well as a Swede and a
Dane. The White House committee was made up of only U.S. citizens.[12]
Larry Regens, of EPA's Office of International Activities, notes, however, that of the nine panelists, four were members of the National
Academy of Sciences and one was a member of the National Academy
of Engineering, both of which are independent of direct government
oversight.[13]

The U.S. review was completed in the spring of 1983, and the panel released its summary of findings in June. The panel acknowledged that there are many unanswered questions concerning acid rain and noted that scientists are uncomfortable with such uncertainty. But it also concluded that political action cannot wait for complete information: "There are many indicators which, taken collectively, lead us to our finding that the phenomena of acid deposition are real and constitute a problem for which solutions should be sought." The panel report goes on to state that acid rain belongs to the class of social problems that cannot be solved once and for all by applying "a straightforward sum of existing technological and legislative fixes." Instead, knowledge about the problems and political action must evolve over time. "It is in the nature of the acid deposition problem that actions have to be taken despite incomplete knowledge." This should be done in such a way that a steady increase in knowledge and understanding can be accommodated, "taking various actions that appear most effective and economical at any given time." With this strategy in mind, the panel report recommends, as examples of first "least-cost" steps, burning fuel of different sulfur content during different seasons, gross reductions in sulfur emissions from nonferrous smelters, and intensifying coal washing.[14]

The tone of the document, in striking contrast to earlier administration statements, was one of concern. The group felt that some of the effects of acid rain may be severe and irreversible. Several committee members were particularly concerned about the effects of acid deposition on the microorganisms that degrade natural wastes in soil, essential for recycling nitrogen and carbon in the food chain. The chairman of the panel stated that with such a "worrisome thing . . . you're not going to sit around and wait for 20 years" to get conclusive proof of the danger.[15]

The Canadian peer review group issued its report in May 1983 and summed up its reading of the available evidence in the following way:

1. Over North America the area of most acid deposition lies over, and downwind from, the major industrial regions on the continent. Acidity is many times greater than natural background levels (up to a hundredfold in the worst areas).

2. The acidity is due to *the presence of sulphur dioxide and nitrogen oxides, and to their conversion in the atmosphere to sulfates and nitrates.*

3. The acid-forming gases are carried from sources to vulnerable areas by the winds. *While in transit they undergo complex chemi*

cal changes, which are not yet fully understood, over considerable distances and long periods of time (such as a year). These
facts are not expected to affect the basic "linearity" of the system:
that is, *to halve the deposition it will be necessary to halve the
emissions.*[16]

The findings of the peer review groups have been reported to the
official negotiators in several meetings between U.S. Secretary of State
George Shultz, Canadian Minister of External Affairs Allen MacEachen,
U.S. EPA Administrator William Ruckelshaus, Canadian Minister of the
Environment John Roberts, and the two chairmen of the peer review
committees, Bill Nierenberg for the United States and Kenneth Hare
for Canada.

The International Task Group of the Interagency Task Force on Acid Precipitation

Independent of the MOI and its work groups, some bilateral research
activities on acid rain are under way as part of a large-scale and long-
term U.S.-initiated field study. The Interagency Task Force (ITF) was
established by the Acid Precipitation Act of 1980 to plan and coordinate the U.S. National Acid Precipitation Assessment Program (NAPAP).
The task force is intended to direct a long-range research effort rather
than an evaluation of existing information. It was begun in parallel
with the MOI work groups, and formal negotiations were taking place
while the work groups were issuing reports, although the negotiations were often overshadowed by the reports. Now that the work
group reports are completed, the ITF is expected to become the center of activity. The administration has allocated increased funds for
ITF activities, thus emphasizing its commitment to a research strategy
for dealing with acid rain.[17] Many of the same scientists and agency
administrators involved with the MOI work groups are members of
the ITF as well.[18]

The ITF includes twelve federal agencies. Its activities were originally organized into ten task groups, with the head of each being the
representative to the Research Coordinating Council (RCC). Nine of
the task groups were assigned to the research areas outlined in NAPAP
and one (the International Activities Task Group) was responsible for
coordinating those research efforts with work being done in other
countries. Since 1985 the research activities have been assigned to
seven task groups coordinated by a director of research, who is
responsible for planning, coordinating, and recommending the annual

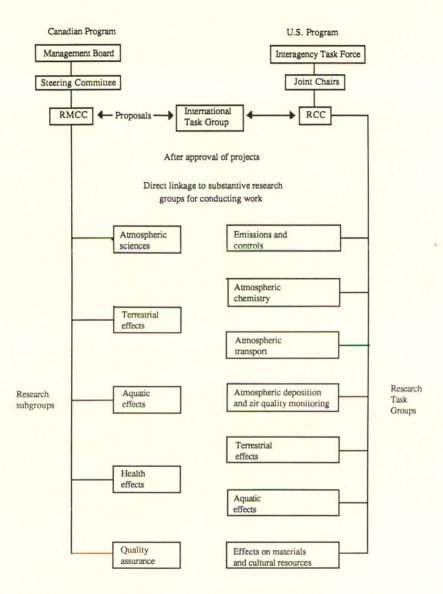

Figure 4-1.   Mechanisms for Coordinating Research between
the U.S. Interagency Task Force and Canadian Research Groups

research program to the joint chairs. The responsibility for managing the assessments was also transferred to the director in 1985.[19]

The International Activities Task Group seeks research coordination and cooperation between the United States and Canada on the acid precipitation issue. In April 1982 the first joint meeting took place between the task group and its Canadian counterpart, the Federal/ Provincial Research and Monitoring Coordinating Committee (RMCC). A mechanism for scientific cooperation was approved, and working-level contacts between scientists of the two nations were encouraged. Specific proposals for coordinated research would be submitted to the RMCC and the RCC of the Interagency Task Force for approval. Approved projects would be channeled to the appropriate task group and its Canadian counterpart under the RMCC for action (see figure 4-1).

During the 1982 meeting the committee reviewed a series of cooperative research proposals and discussed the progress and problems of nine ongoing projects. As a result of the meeting, the two governments agreed to increase cooperative efforts on quality assurance, especially in precipitation monitoring.[20]

In the fall of 1983 Canada and the United States also agreed on a joint monitoring project to study dispersion. Under the Cross-Appalachian Tracer Experiment a harmless and easily measurable tracer gas was released into the atmosphere from several locations in the United States and Canada. Two release stations—Dayton, Ohio, and Sudbury, Ontario—were chosen because they are located close to major pollutant sources affecting the air quality over the northeastern part of the continent. Preliminary analysis of the results indicated that the tracer plume traveled over the ground sampling stations at heights between 300 and 1,200 kilometers (as forecast) and aircraft samples detected the plume from 200 to 900 kilometers downwind of the release points.[21]

Cooperative research between the United States and Canada continues to be the official policy of both countries. The ITF and RMCC meet on a regular basis to discuss progress and plan joint projects. At the scientific level there is little disagreement between the two groups. However, this is not always true when the discussion spills over into policy implications, and sometimes this results in disagreements over the priority that should be assigned to different projects. Because each organization acts in a coordinating capacity, the research is conducted by a variety of federal and state/provincial agencies. Table 4-4 lists the active projects as of December 1987.

The Trilateral Scientific Committee on Acid Precipitation

The Trilateral Scientific Committee is the first coordinated international research effort undertaken by the National Academy of Sciences (NAS). The idea for the committee was born when the presidents of the National Academy of Sciences and the Royal Scientific Society of Canada (RSC) met by chance at an airport. The Canadian president expressed envy at the role of the NAS as an independent adviser to the government (the RSC has a comparatively limited role in Canada). They decided to undertake a joint project on acid rain that would help the RSC gain experience with producing advisory reports.

The Joint NAS/RSC Scientific Committee met once in 1980. Members discussed the U.S. National Research Plan, the Canadian National Research Plan, and the MOI work group documents. The general sentiment was that the MOI groups had not researched the data thoroughly. The Mexican Academy of Sciences (MAS) was invited to join the committee in June 1982. The MAS's interest in acid precipitation stems from a concern about air quality in Mexico City, one of the most polluted urban areas in the world. Increased lignite use in Texas could also affect Mexican air quality.

At the request of the Reagan administration, the Trilateral Committee did not meet for the next six months in 1982 because the peer review of the final MOI work groups was nearing completion. The committee did meet for two days in Washington in December 1982. The meeting was intended to begin the first stages of a new research effort by the committee. Priorities were identified, such as formalizing monitoring and research between Canada and the United States, and identification of gaps in research. A trend assessment, using existing data from the past two decades, will be a primary focus.

Three focuses for study were identified at the meeting:

1. Do existing data reveal consistent patterns (precipitation chemistry, surface H_2O, lake sediment, tree rings)? This study was to be administered by the NAS.
2. How much deposition can a given ecosystem sustain without significant changes?
3. What kinds of institutional mechanisms are needed to deal with acid precipitation?

Securing funding for these activities turned out to be difficult. Some internal NAS money was made available for the project. It was hoped that the EPA would finance part of the research, but then-Administrator

Table 4-4.   U.S.-Canadian Cooperative Acid Deposition
Research Projects

Activity	Purpose
Forest Effects	
North American Forestry Commission	Acid Deposition Study Group: to coordinate research on forest effects/atmospheric deposition in U.S., Canada, and Mexico
Data exchange	Impact of acidic fog and clouds on forests
Ongoing research	Assess causes of sugar maple decline in north-eastern U.S. and Canada
	Decline of western conifers
Aquatic Effects	
Monitoring program	Exchange of information on protocols and quality assurance activities
Data exchange	Interpretation and evaluation of U.S. and Canadian surface water chemistry surveys
Atmospheric Projects	
Joint research planning and design	Eulerian acid deposition and oxidants model development
Field study	Data collection for performance evaluation of acid deposition models
Model evaluation	Evaluation of Eulerian acidic deposition model
Cooperative field study	Determination of pollutant concentration, deposition, and horizontal flux with distance over the North Atlantic (WATOX)
Data exchange	National emissions inventories
Equipment transfer	Air quality/fog sampling at the Bay of Fundy and other eastern North American sites
Depositon Monitoring	
Network inter-comparison, co-location of samplers	Intercomparison of wet deposition networks, i.e., U.S. National Trends Network (NTN) and Canadian Air and Precipitation Monitoring Network (CAPMON)
Laboratory intercomparison	Intercomparison of chemical methods used to analyze wet deposition samples
Common data base	Unified acid deposition data base for eastern North America

| Participating Organizations | |
U.S.	Canada
USDA Forest Service	Forestry Service
USDA Forest Service, EPA, State University of New York, Albany	Forestry Service, Atmospheric Environment Service, McGill University
USDA Forest Service	Forestry Service
USDA Forest Service	
EPA	Environmental Conservation Service
EPA	National Water Research Institute
EPA, State University of New York, Albany	Ontario Ministry of the Environment, Atmospheric Environment Service
EPA, EPRI NOAA, TVA	Ontario Ministry of the Environment, Atmospheric Environment Service
EPA, EPRI	Atmospheric Environment Service
University of Virginia, NOAA, Bermuda Biological Station	Atmospheric Environment Service
EPA, DOE, NOAA	Environment Canada
NOAA	Environment Canada
USGS, National Atmospheric Deposition Program, Illinois State Water Survey	Atmospheric Environment Service, Environmental Conservation Service
USGS, Illinois State Water Survey	Atmospheric Environment Service
EPA, USGS, National Deposition Program, Pacific NW National Laboratory	Ontario Ministry of the Environment, Atmospheric Environment Service

Table 4-4.   (continued)

Activity	Purpose
Methods research	Development of standard methods to measure dry deposition
Network inter-comparison co-location of samplers	Intercomparison and co-location of sites for planned dry deposition monitoring networks
Monitoring	Continue to develop and expand inventory of wet deposition monitoring sites within transboundary region
Co-location of samples at a variety of sites	Effects of acid deposition on the corrosion of structural materials
Other Joint peer reviews	Canadian participation in task group research peer reviews

Source: EPA, mimeo, December 1987.

Gorsuch declined, saying the studies were duplicating already completed work.[22]

The trends study is under way. The subcommittee directing the study is administered by the NAS and includes some fifteen members. The committee agrees that the effort to provide the Canadian and Mexican academies with practical experience in developing independent, objective reports to provide a base for decision making is as important as the actual research. MAS and RSC members are observing and participating in the steps of selecting members, carrying out studies, ensuring public access, and conducting peer reviews.

When it was learned that the MOI work group reports would undergo formal peer review, the Joint NAS/RSC Scientific Committee offered to do the review, but, as noted earlier, the Reagan administration preferred a unilateral review administered from the White House. The administration may have opposed the appointment of the Joint Scientific Committee because it considered the work undertaken by the NAS to be supportive of the Canadian position. A 1981 report by the NAS, *Atmosphere-Biosphere Interactions*, had recommended a reduction in hydrogen ion deposition of 50 percent.[23] The report suggested that deposits of acid compounds would have to be cut by 50 percent

Participating Organizations U.S.	Canada
NOAA, USGS	Atmospheric Environment Service
NOAA, EPA	Atmospheric Environment Service, Environmental Conservation Service
EPA, NOAA, DOE	Atmospheric Environment Service
Bureau of Mines, EPA, National Park Service	Atmospheric Environment Service
NAPAP, Office of the Director of Research	RMCC/LRTAP Liaison Office

to avoid further damage to the most sensitive lakes and streams in the northeastern United States and southeastern Canada. The press wrongly interpreted this as a recommendation for a 50 percent reduction in emissions of SO_2 and NO_x. It was further assumed that this would result in a parallel reduction in acid deposition. But the question of how much a given reduction in emissions would yield in reduced damage had not been addressed by the report.

In response to the misinterpretation of the 1981 report, another study was conducted to help clarify the relationship between emissions and deposition. The NAS's Environmental Studies Board established the Committee on Atmospheric Transport and Chemical Transformation in Acid Precipitation to try to answer this question. Their report, *Acid Deposition: Atmospheric Processes in Eastern North America*, was released in June 1983.[24] The committee concluded that the relationship between emissions and deposition was indeed linear: a given percentage reduction in emissions of sulfur dioxide would result in a corresponding reduction of acid deposition. The occurrence of acid rain in eastern North America appears to be roughly proportional to the average annual emissions of sulfur dioxide from power plants, industrial facilities, and other sources in that region.

This was a significant finding. But the committee also added that scientific models describing the movement of acid-forming pollutants over long distances were not yet precise enough to analyze the effects of emission control strategies.

In addition to clarifying the linkage between emissions and deposition, the report was intended to identify research priorities for regulatory development. The final chapter of the report is devoted entirely to this task, and the following priorities were defined:

1. *Dispersion, chemical transformation, and transport.* Precision models are probably a long way away, but useful data about transport of acid to rural areas could be gathered quickly by empirical observation. Studies of cloud processes and the rate and seasonal variation in the chemical transformation of SO_2 and NO_x are needed.
2. *Dry deposition.* There is a need to develop an accurate methodology for measuring dry deposition, which may account for as much as half of all acidic deposition.
3. *Tracers.* Methods need to be developed for tracking particular air parcels on a regional scale. The impact of specific source regions upon specific receptor regions could be studied if a method for tracing existed.

The report was received as an important contribution to the study of acid deposition. Most of the first printing of 3,500 copies was sold within a few weeks. Interest in the report has been described as higher and more sustained than for most NAS studies. A copy of the report was delivered to the president in late September 1983, and it was discussed by the cabinet. The National Governors' Conference has taken note of the report and established a task force on acid rain to prepare a policy statement. On the international level, the Organization of Science Counselors in Washington, D.C. (consisting of the science counselors from forty-five embassies), requested a formal presentation of the main findings of the NAS report. The White House Office of Science and Technology Policy was careful to release its own acid rain statement several days prior to release of the NAS's study. In late July the NAS conducted a briefing on its report for the administrator of the Environmental Protection Agency and the presidential science adviser.

State-Provincial Agreements

States and provinces along the U.S.-Canadian border are in constant contact on a great variety of issues of mutual concern. Because many of the U.S. border states are as concerned as the Canadian provinces about acid deposition, both formal and informal meetings between government officials are common. In 1985 the governors of the New England states and the eastern Canadian provinces agreed to develop plans to reduce SO_2 emissions. However, no formal agreements for reduction targets have been reached and all of the other agreements relate to monitoring activities and the exchange of data.[25]

For instance, New York and Quebec have agreed to use standardized methods and procedures in analyzing samples from existing collection sites and to publish an annual report on the quantity and quality of acid deposition. Research activities will be coordinated and in some cases joint studies will be undertaken. Similar agreements have been reached by Wisconsin and Quebec, Michigan and Ontario, Minnesota and Ontario, and New York and Ontario.

Conclusions

Joint scientific research groups can play an important role in the development and interpretation of the scientific evidence needed to resolve transboundary problems. This role is amplified when an issue is surrounded by scientific uncertainty and disagreement. In such a situation research groups can be vital as assessors of mutually accepted scientific evidence and as advisers to policymakers. Even if an agreement is not reached, they can point out to policymakers the nature and extent of disagreement. Earlier environmental disputes between Canada and the United States used advisory boards assembled by the International Joint Commission for the Great Lakes to assess the state of knowledge.[26] In the case of acid rain the decision was made to assemble bilateral research groups as part of the formal negotiating process. While the experiment was marred by problems related to the political transition between two administrations in the United States, it nevertheless represents an important first in bilateral relations because the contributions of scientific expert groups were made a central component of the diplomatic process.

In retrospect, a number of questions can be asked, not so much to criticize what was done but to be taken into account in future cases: Was is wise to select members of work groups primarily from among scientists working for the two governments? Was the scientific community (NAS, professional associations) sufficiently prepared to offer structured scientific advice? Given the lead of Scandinavian scientists in analyzing acid rain, was their experience sufficiently taken into account? Would a standing mechanism for bilateral scientific advice, rather than ad hoc groups, provide better, faster, less politicized advice? On the other hand, involving the science adviser to the president or prime minister at some point during the proceedings appears to make good sense. Both sides, perhaps, did surprisingly little to bring into play those individuals and groups who are most experienced in bridging the gap between science and politics. Why? And what can be learned from this apparent failure to heed a top-level advisory mechanism that, at least in the case of the United States, has tried for more than two decades to bring the sober voice of science to policy-making?

‖‖‖‖‖

‖‖‖‖‖

Part Two

Domestic Policy Development

‖‖‖‖‖

Introduction to Part Two

Although acid rain is quintessentially an international problem, we maintain that domestic policies for acid rain control must be in place before significant progress will be made in bilateral arrangements between Canada and the United States. This seems to go against the logic of the acid rain problem. The entire northeastern part of the North American continent is affected (and other areas may become affected in the future), reflecting geological conditions in the region as well as concentration of industrial activities. A full-fledged bilateral control program, therefore, would seem superior from the point of view of efficiency. Politically, however, this seems unrealistic. Given the nature of the political process, domestic solutions will have to be in place before international activities of much consequence can follow. The dominance of domestic policy development does not exclude useful but limited international programs that could be implemented in the near future. This would include joint research programs, monitoring of emissions and deposition, and consultations among experts as well as policymakers.

If domestic action must lead the way, what does this mean for the two countries? We suggest that without meaningful action on the part of the United States, given its larger population and economy, little progress can be made in controlling acid rain in North America. The debate in the United States has focused on the reauthorization of the Clean Air Act. Numerous groups are involved in the process, ranging from traditional business lobbies, to broad-based public interest groups, to affected government agencies at federal, state, and local levels. Three types of bills are under consideration: those calling for more research, those calling for pollution-loading limitations, and those calling for more comprehensive controls of transboundary pollution.

In the case of Canada, the federal government can take the lead in international matters but is severely constrained in its powers in domestic policy development. However, there is much more consensus in Canada concerning the urgency of the acid rain problem. Consequently, the federal and provincial governments have been able to agree on an arrangement that allocates emissions reductions between the seven eastern provinces. Most of these provinces now have programs in place that will result in an overall reduction in SO_2 emissions of 50 percent by 1994.

The chapters in this part of the book explore the initiatives currently under way on both sides of the border to develop a national acid rain control strategy. The legal and political constraints on both countries and the roles of the various interest groups in the policy process are detailed. The extent to which a national solution is possible and probable is examined. The different federal structures and politics of the United States and Canada have produced very different approaches to acid rain, and it becomes clear why the attention in the United States has focused on the reauthorization of the Clean Air Act, and in Canada on exerting pressure on the United States. As a result, different interest group alignments have emerged—in the United States the forces for and against control can be found in the traditional camps; but in Canada industry and environmental groups are united in their call for strong measures against acid rain. It is important to understand these differences because they will determine the outcome of the policy debate in the two countries.

Chapter Five
Environmental and Economic Interests
in Canada and the United States

||||||||

Introduction

It is apparent from the discussion in previous chapters that acid rain is a complex problem without a ready solution. Even with the best of intentions, eliminating acid deposition will not be easy given the high level of uncertainty associated with assigning cause-and-effect relationships. Added to this, the costs associated with controls could be very high and not necessarily distributed equitably. It is therefore not surprising that many interest groups in both the United States and Canada have been involved in the debate. This chapter examines some of the sources of the controversy and how these interest groups influence the arguments for and against the control of acid precursor emissions.

The positions of some representative groups are characterized by the groups' perceptions of four distinct arguments embodied in the controversy: the scientific argument over the causes, effects, and properties of acid rain;[1] the ethical argument over the appropriate action in the face of uncertainty; the economic argument over the costs of control and damages; and the political argument over the allocation of authority and responsibility among groups and government units.

In general, proponents and opponents of controls can be identified as victims and polluters, respectively. The argument appears to center on the issue of who should bear the costs of controls: those who pollute or those who have the most to gain (that is, the current victims). In the United States this has resulted in the usual adversarial

---

This chapter was written by Tom Albin and Steve Paulson.

relationship between industry and environmentalists. However, in Canada, which increasingly sees itself as a victim of U.S. pollution, a more unified attitude has emerged.

Obstacles to Acid Rain Policy

The Scientific Argument

Environmental policy almost always involves action in the face of uncertainty. The effect of scientific uncertainties on policy formation is aggravated in the case of acid rain because so much is at stake and because the costs and benefits of control measures fall unevenly across political subdivisions. Those who propose controls argue that we now have enough evidence to justify action because the potential effects are so devastating; they argue, in fact, that we have more evidence than we had when imposing past environmental legislation. Opponents of controls argue that the evidence is insufficient to ensure the selection of an efficient control strategy and that past environmental policies have been unnecessarily costly and burdensome because they were formulated in haste.

Policymakers therefore face a dilemma in constructing an acid rain policy: If they fail to act, they risk responsibility for widespread and irreversible damage to the environment; if they act on the basis of insufficient scientific knowledge, they risk deepening unnecessarily the economic woes of a nation. It is important to note, however, that the research undertaken so far is slowly confirming original theories about the transport and transformation of SO_2 and NO_x in acid deposition.

The real scientific obstacles to acid rain policy are the competing theories about the ability of the environment to absorb pollutants and neutralize acids, and the poorly understood roles of each of the possible acid rain precursors. Scientists do not know whether the ability of the environment to buffer acids is continuous or if there is a threshold for absorbing pollutants beyond which buffering is impossible. If the environment is capable of buffering a certain amount of acid continuously with limited detrimental effects, then a strategy restricting emissions levels to ensure that this buffering capacity is not outrun would be sufficient. However, even if this theory is correct, policymakers might well decide that some areas where the buffering capacities are lowest might not be worth the cost of even this level of control. If the second theory is correct, and there is an absolute threshold for absorbing pollutants, then the problem becomes much more seri-

Table 5-1.   U.S. Sulfur and Nitrogen Oxide Emissions by Source, 1980

Source[1]	Sulfur oxides		Nitrogen oxides	
	Million metric tons/year	Percentage of total	Million metric tons/year	Percentage of total
Mobile sources	0.9	3.8	9.1	44.0
Highway vehicles	0.4	1.6	6.6	31.9
Aircraft	0.0	0.0	0.1	0.4
Railroads	0.1	0.0	0.7	3.2
Vessels	0.3	1.2	0.2	0.8
Off-highway vehicles	0.1	.0	1.5	6.3
Stationary sources	19.0	80.0	10.6	51.0
Electric utilities	15.9	67.0	6.7	32.0
Industrial	2.3	9.7	3.3	15.9
Commercial-institutional	.6	2.5	0.3	1.2
Residential	.2	.8	0.3	1.2
Industrial processes	3.8	16.0	0.7	2.9
Total amount emitted	23.7		20.7	

1. Sources are not mutually exclusive.
Source: Council on Environmental Quality, *Environmental Quality: 12th Annual Report* (Washington, D.C., 1981), p. 246.

ous because the important factor becomes the total accumulation of acids deposited in any given area. In this case the problem becomes one of restricting the loadings of emissions as severely as possible because deterioration will continue as long as *any* acids are being deposited. Under either theory some type of regional regulation is required to address the problem; the question that the difference in these theories raises is how severe should the regulation be?

Initially the debate centered on the control of SO_2 emissions from stationary sources. Table 5-1 indicates that this is one logical focus for the debate, because these emissions account for 19 million metric tons of acid rain precursors out of a total of 44.4 million metric tons emitted in 1980. The control of SO_2 emissions under existing statutes has resulted in a decrease of 27 percent in ambient levels between 1973 and 1985. Between 1977 and 1985 SO_2 emissions from coal-fired plants decreased 11 percent, despite a 40 percent increase in coal use.[2] This is primarily because sources are burning cleaner fuels, locating away from cities, and, until recently, were building taller stacks to disperse emissions over larger areas. Technology is available for controlling at least 90 percent of SO_2 emissions from coal-fired power plants; however, chemical scrubbers for smokestack exhausts are expensive, and only sources built since 1978 are required by federal law to use them.[3] Older power plants are allowed to emit five to seven times more SO_2 than new sources.[4]

NO_x emissions from mobile sources may ultimately be the more difficult pollutant to control. NO_x emissions are now rising while SO_2 emissions are falling, and technology seems to promise more opportunities for further reducing SO_2 emissions than for reducing NO_x emissions. Some theories also suggest that NO_x concentrations build up in snow to higher levels than do SO_2 concentrations and thus cause more severe acid-shock effects during spring thaws. Despite current controls aimed to reduce suspected negative health effects, NO_x emissions have risen in recent years and are expected to rise further due to increases in miles traveled and electricity generated. Highway vehicles contribute 31.9 percent of total NO_x emissions, and electric utilities contribute 32 percent, so these sources have become targets for controls.

Another source of scientific uncertainty is the role of other atmospheric components. Increasingly, hydrocarbons and ozone are being viewed as important air pollutants. In the acid rain story they would appear to contribute to acid deposition by acting as precursors in the formation of hydroxyl radicals, which are extremely effective oxidation agents for the conversion of SO_2 to sulfate or sulfuric acid.[5]

Science must also answer questions concerning the reversibility of acid rain effects, the feasibility of mitigation strategies as a complement to or substitute for control measures, and the prospects for controlling the conditions under which acid rain forms. These questions all enter into the debate at some point, as interest groups seek to promote or forestall control measures and politicians seek solutions that provide effective protection of environmental resources at the lowest possible costs.

The Ethical Argument

The ethical argument arises directly from the uncertainty of the scientific argument.[6] On one side, the public and private interest groups promoting the imposition of some control strategy for the apparent precursors to acid rain maintain that, in the face of uncertainty about environmental effects of emissions, it is better to err on the side of the environment than on the side of short-term economic benefit. These groups promote action to protect the biosphere on the grounds of an ethical responsibility to present and future generations, even if the benefits of that action are uncertain and the costs great. On the other side of this argument are various groups that oppose the imposition of control strategies because these strategies cost too much—in dollars and in disruption of lives—and may accomplish little or nothing. These groups oppose controls, without greater evidence that favorable results will occur, on the grounds of an ethical responsibility to protect ratepayers, investors, and miners from high costs and disrupted lives.

The Economic Argument

The economic argument concerns the magnitude of the costs of control or of inaction, and who should bear these costs. Both the costs of inaction and of action to control emissions are subject to a wide range of speculation. Some proponents of controls give little attention to the question of who should bear the costs, just as many opponents of controls, who usually stand to lose economically from regulation, give little attention to the costs of inaction. Again, uncertainty is the rule in determining costs, and some groups on both sides of the argument use methods and figures that suit their interests, even when those methods and figures are demonstrably faulty.[7] On the other hand, it is even more difficult to quantify the benefits. Environmental benefits are notoriously difficult to evaluate in monetary terms, and given that the exact relationship between emissions controls and reductions in acid deposition has not been established, it may be many years before the advantages of controls are known.

In the final analysis, the economic impacts are likely to determine the outcome of pending legislation, and there are costs associated with inaction as well as action. The issue is not whether any emissions controls are necessary, but how stringent they should be. This is particularly pertinent because, as with most pollution control technology, costs do not increase in a linear manner. For instance, the Office of Technology Assessment (OTA) estimates that five billion tons of utilities' SO_2 could be eliminated for less than $1 billion, whereas it would cost $4 billion to eliminate ten to eleven million tons.[8] The range of issues includes:

1. *Agricultural damage.* There is little evidence of damage to crops from acid deposition; the major source of damage is ozone.
2. *Damage to forests.* Until recently the most spectacular evidence of forest damage came from Europe, but now there is mounting evidence of forest death in the eastern United States and Canada. In Canada "maple die back" is threatening the maple syrup industry, worth $40 million annually. Approximately one-half of Canada's productive forests, which generated $14 billion worth of forest products in 1982, are receiving high levels of acid rain.[9]
3. *Structural damage.* Although the extent of damage to bridges, buildings, and monuments as a direct result of acid rain cannot be quantified, there is evidence that it is considerable. Estimates range from $15 million to $60 million.[10]
4. *Health effects.* SO_2 is a powerful lung irritant and may be the third leading cause of lung disease. The congressional OTA determined that airborne sulfur particulates at 1978 emission levels may be responsible for 50,000 premature deaths.[11] However, it is not clear whether there is a direct relationship with acid deposition.
5. *Commercial and recreational fishing.* Acidification of lakes can be directly correlated with fish death, particularly below pH 5.[12] Forty-three percent of the lakes in Quebec and Ontario are vulnerable to acidification; about 90,000 jobs are at risk in the eastern Canadian commercial fisheries and 600 fishing camps may be closed by the year 2000.[13]
6. *Cost increases for the utility industry.* The OTA estimates the total cost of reducing SO_2 emissions by ten million tons annually as $3–$4 billion, the equivalent of a nationwide average electricity rate increase of 2–3 percent for residential customers.[14]
7. *Cost increases for new automobiles.* Estimates for limiting NO_x emissions of new cars to 0.7 grams per mile traveled vary from $27 to $99 per car.[15]
8. *Unemployment in the coal industry.* Emissions reductions by unspecified means would favor the production of low-sulfur, western coal, and could cause the loss of 20,000–30,000 mining jobs (for a reduction of ten million tons of SO_2). Legislation that mandates high-removal control technology could minimize such dislocations, but would increase total program costs by about $1.5 billion annually.[16]

The Political Argument

Acid rain by its very nature requires different political handling than environmental problems usually receive. In recent years environmental policy increasingly has been treated as an aspect of health policy, and environmental legislation is usually justified on the basis of human health effects. Thus, control of toxic chemicals and nuclear waste disposal are treated as more pressing environmental problems than acid rain because these problems apparently pose a greater threat to human health. Acid rain may pose not yet fully understood health threats, but its main threat is to the environment at the ecosystem level. It does not cause dread diseases in humans; instead, it threatens the ability of the natural environment to renew itself. For these reasons acid rain requires the formation of a comprehensive environmental policy unlike any now in place.

Besides its concern mainly with immediate human health effects, past environmental policy has been inadequate for controlling acid rain as a large-scale regional problem. Policymakers usually impose uniform national standards or allow states to impose statewide standards. But acid rain does not recognize political boundaries. Since the most concentrated emissions of acid rain precursors occur in the Ohio valley, and the most highly sensitive regions are in the northeast and north-central United States and Canada, the imposition of uniform national standards would benefit certain regions at the cost of others, without any particular attention to the equity or the efficiency of such measures. Furthermore, emissions originating in some areas clearly are more dangerous than those originating in others. Thus, state regulation is inadequate, but uniform national standards would be both inequitable and inefficient.

At this point it becomes evident that the political and economic arguments are almost inseparable. The Congressional Budget Office has analyzed the cost of four approaches for financing and administering emission controls: (1) the polluter pays approach: utilities and their customers pay for the control; substitution of low-sulfur coal is not an option; (2) scrubber installation is partially financed by an electricity tax; (3) sulfur dioxide emissions are taxed; partial subsidies for installing scrubbers; (4) sulfur content of fuel is taxed; partial subsidies for installing scrubbers.

The use of subsidies would transfer the costs from those who pollute the most to those who pollute the least. In choosing between these alternatives, the objective is to strike a balance between competing goals—achieving emissions reductions, alleviating costs to consumers and utilities in any one region, and minimizing disruptions to the coal industry.[17]

The "polluter pays" principle, usually accepted in the past, is simply unrealistic in this case, for several reasons. One is the geographic distribution of the costs and benefits mentioned above. The polluting states are already suffering from high unemployment and loss of industry to other locations. They will suffer further economic impacts if strict regulations are imposed without some way to distribute costs of regulation. Furthermore, there is no way to force the polluter to pay. Electric utilities can (and must) pass on the costs of complying with regulations to consumers. Regulatory measures that would result in regional increases in utility rates are unlikely to gain easy acceptance. The situation is further complicated by the division between low-sulfur and high-sulfur coal producers. Regulations requiring the installation of scrubbers would favor high-sulfur coal producers, while more flexible regulations would allow utilities to switch to low-sulfur fuels to reduce emissions.

The regional aspect of the debate, in which the Northeast stands to gain the most and the Midwest stands to lose the most, is critical in resolving the issue of controlling acid rain. The significance of this issue changes at different levels of control. As explained in the previous section, the marginal cost of controls increases at an alarming rate. At a reduction level of twelve million tons of SO_2 it approaches $2,775 per ton, because the use of scrubbers would be mandatory; at lower levels switching to low-sulfur coal is a relatively cost-effective option.[18] These economic issues bring several politically powerful interest groups into the process and make policy formation more difficult. Any policy must address specifically the questions of who must pay for controls and how costs will be assessed.

The question then becomes one of deciding whether cost allocations should emphasize equity or efficiency. Over the last eight years a great many bills dealing with this subject have been introduced into the U.S. Congress. The costs associated with some of these measures have been analyzed by David Streets.[19] For any given target reduction level, the bills that most closely approach the least-cost curve are those with the most flexibility. Measures that allocate reduction levels by state, or that require the installation of scrubbers, increase costs. It is apparent from the bills that have been introduced that the trend over time has been to accommodate regional interests, thereby sacrificing economic efficiency. This trend has been driven by the realization that powerful interest groups can prevent the enactment of legislation that is against their best interests. However, it can be argued that economic efficiency is not necessarily more valid than political feasibility if, in fact, the latter reflects a greater range of costs and benefits. The number of parameters that can be quantified is always limited, and social costs associated with the disruption of communities is not one of them.

The political argument also involves the allocation of authority and responsibility among groups and governmental units. This aspect of the debate will determine how the scientific and economic uncertainties will be handled. Consequently, it most directly determines action or inaction. Many questions are involved in the political argument over acid rain. For example, what research organizations, government agencies, or groups will officially investigate the scientific and economic uncertainties? To what extent and in what ways will the two federal governments cooperate with each other and with other levels of government? What responsibilities will the two nations recognize toward each other? What responsibilities will the levels and branches of government within each country recognize toward each other? How will the "losers" from adopted policies be compensated?

The Interplay of the Four Arguments

Certainly the scientific, ethical, economic, and political arguments overlap, and involved groups' fundamental stances on one argument can determine their stances on the others. Many groups that advocate controls, for example, give little attention to the economic argument, assuming that the scientific and ethical arguments should determine the course of political action. Conversely, utilities' interests effectively ignore the important elements of environmentalists' ethical position by relying on cost-benefit analyses that do not begin to specify all of the costs and benefits. However, both groups must realize that any successful policy response must deal explicitly with each argument.

These four arguments play very different roles in the policy debates in the United States and in Canada. The U.S. debate focuses sharply on the reauthorization of the Clean Air Act and the amendments proposed to deal specifically with acid rain. The battle lines are clearly drawn along fairly predictable lines: environmentalist groups—along with certain health, research, and tourism interests—strongly support passage of regulatory measures for SO_2 and NO_x, while coal and utility industry groups claim that uncertainty is too great to impose regulation and that more research is needed. The Reagan administration sides with the utilities on all of the fundamental arguments.

In Canada, however, the disposition of regulatory authority and the widespread perception that Canada is the victim of U.S. pollution has brought about a different alignment of interests. Canadians realize that acid deposition does originate from industrial sources, many of them in the United States. The major element of Canadian federal policy is to encourage action by the provinces to control stationary sources of pollution and to negotiate an agreement with the United States to reduce its levels of emissions. This situation has modified the roles of public interest groups, which have concentrated on increasing public awareness both in Canada and the United States. Furthermore, given the widespread public sentiment in Canada against acid rain, and the perception that U.S. industry is largely to blame, some major Canadian industries have conceded many of the environmentalist claims that U.S. industry is still debating. Thus, Canadian industries have joined the Canadian government and environmental groups in attempting to bring about stricter regulation in the United States. This situation has led to some speculation in the United States about the existence of a "Canadian conspiracy" to open the U.S. electric power industry to Canadian competition.[20]

Interest Groups' Positions on Reauthorization of the Clean Air Act

The focal points of debate in the United States are the bills to amend the Clean Air Act (see table 7-2).[21] These bills are of two types: bills funding more research (supported by the utility and coal industries and the Reagan administration as an alternative to controls) and bills calling for pollution-loadings reductions. The latter type have generated the most controversy. Many bills have been introduced; each differs from the others in the specific provisions, but each requires eight- to twelve-million-ton reductions over a specified period of years (ten to fifteen years in most cases), using 1980 as the base case. Initially the debate between interest groups was polarized. In general, with the noted exception of the Tennessee Valley Authority, most interest groups actively involved in the issue are either proponents or opponents of strengthening the Clean Air Act and providing for acid rain controls. They could be loosely defined as either environmental or industrial in nature. More recently, with increased support for strengthening provisions, the debate has focused on the nature of the control measures to be adopted. As a result, industry groups are representing a wider range of interests, depending on how the economics of specific provisions affect them individually.

With attempts to reauthorize the Clean Air Act deadlocked for so long, some states, particularly those in the Northeast, have come to realize that initiatives at the state level are appropriate and timely. Because most states are able to control less than half of the acid deposition they receive, a cooperative approach is likely to be the most productive. On the other hand, the states as a whole represent the same spectrum of interests as the interest groups. In an attempt to break the deadlock, a group of governors, headed by Tony Earl of Wisconsin and John Sununu of New Hampshire, created a forum in October 1985 to advance the air pollution debate beyond traditionally polarized perspectives. Renamed the Center for Clean Air Policy, it is discussed in a separate section below.

U.S. Environmentalists

National Clean Air Coalition   The National Clean Air Coalition (NCAC) is examined as representative of the "environmental" position in the United States. (The term "environmental" is used here to identify the position of groups that support more stringent control laws.) The problem of acid rain has caused other groups that are directly affected to join with traditional environmentalists. These groups consist of health-care groups, cottage owners, fishermen, and outdoor recreation industry associations who believe that acid rain poses a threat to human health or who could lose their recreational enjoyment or their livelihoods if acid rain is not abated. NCAC presents a position common to environmentalists, health groups, and the victim industries.[22] It asserts that the benefits of controls outweigh the costs, that the costs of controlling acid rain are reasonable, and that control will not cause U.S. coal production to decline significantly.[23] Scientific and ethical concerns primarily determine NCAC's position, but most arguments are about the economics of controls.

According to NCAC, the benefits of a control program would be large: less damage to lakes and fish; lower risks to forests; improved visibility; reduced corrosion of buildings, monuments, and water pipes; and a margin of safety against poorly understood health risks.[24] Since many of the suspected dangers involve long-term, subtle changes in the biosphere, dollar figures cannot begin to estimate damages appropriately. Yet even the partial cost estimates of acid rain damages are large. One study estimates $5 billion per year in damages in the eastern United States alone.[25]

The environmental groups represented by NCAC are in full agreement on the acid rain issue. They concede to opponents of a legislative program to control acid rain that gaps in knowledge of atmospheric chemistry, pollutant transport, and acid rain do exist. However, NCAC's claim of a cause-and-effect relationship between industrial emissions and acid rain is supported by many independent scientists who have studied the problem.[26] According to the National Wildlife Federation (NWF), a member of NCAC: "The vast majority of knowledgeable scientists recognize that, despite the lack of complete information, we know enough about the causes, behavior, effects, and cures of acid rain to not only warrant, but compel immediate action to control the problem."[27]

NCAC is critical of cost-benefit studies by the utility industry, such as those done by the American Electric Power Service Company (AEP) and the Edison Electric Institute (EEI), which assert that pollution controls would raise utility rates astronomically while offering uncertain benefits. It charges that the AEP and EEI analyses exaggerate impacts of controls on the utility industry by ignoring tradeoffs between states, inflating the price of coal, and overpricing the costs of emissions control devices. NCAC members call these distortions a part of the "informational haze" surrounding the utility industry's response.[28] Another element of the informational haze is the utility industry's suggestion that the problem is a Canadian conspiracy, designed to give Canadian electrical power exporters an unfair competitive advantage over U.S. utilities.[29] According to NCAC, this argument, along with scientific uncertainty, is used to "generate a smokescreen, hiding from public view the virtual unanimity of the knowledgeable and independent scientific community about the causes of acid rain. . . . In this connection, it is important to recall where the acid rain issue began —not in a political speech or scare headline, but in the scientific community."[30] On the other hand, an ICF, Inc., study commissioned by the NCAC and NWF was criticized for significantly underestimating the costs of compliance.[31] Differences between ICF and industry cost estimates are discussed more fully in the section on the Center for Clean Air Policy.

NCAC's political activities include testifying before congressional committees, informing legislators of the merits of relevant positions, and, most important, grass-roots public education. Public education is accomplished by furnishing member groups and individuals with analyses of pending bills and by promoting media coverage.

U.S. Industry

The U.S. "industry" groups discussed here include the Electric Power Research Institute (EPRI), the Edison Electric Institute, the National Coal Association (NCA), and the American Electric Power Service Company. (Again, the term "industry" is loosely used to designate those industries perceived to be polluters and opponents of the SO_2 control measures under consideration.) This list, while not exhaustive, is representative. The United Mine Workers Union (UMW), for example, is involved in the debate and could represent a powerful political force, but their position is not substantially different from that of the NCA. The Tennessee Valley Authority (TVA) is discussed as a "quasi industry" group since TVA is a public enterprise that works both as a utilities provider and a guardian of resources.

The utility and coal industry groups involved in the debate take several approaches to the problem. For example, EPRI, as a research organization, generally presents its findings in more objective, less politically charged terms than does EEI, which is primarily a political lobbying organization. EEI, AEP, NCA, and UMW discourage action on the basis of scientific uncertainty, but these groups emphasize much more heavily the economic and political arguments. Through various cost-benefit analyses, these groups argue that costs of electric power will rise significantly, that dislocation in the already troubled coal industry will be aggravated, and that the costs of regulation will fall inequitably to certain regions and industries. NCA and EEI both promote public education on the merits of their arguments, but their efforts on this front have not been as successful as those of the NCAC. The industry groups are generally more successful at lobbying legislators than educating the public.

As an electricity producer, TVA shares many concerns with the privately owned utilities. TVA, however, has conceded many of the environmentalists' claims about the causes and effects of acid rain. The recent experiences of TVA in installing control equipment and promoting conservation suggest that utility industry cost-benefit analyses overestimate the costs of regulation and that environmental groups underestimate these costs. Although TVA has not taken an explicit position on whether controls should be imposed, spokesmen have suggested that significant emissions reductions are possible at moderate costs.

More recently, as support for strengthening provisions has increased, the debate has focused on the nature of the measures to be adopted. As a result, industry groups are not necessarily speaking with one voice. Several new lobbying groups have formed, some with the specific intent of protecting the interests of low-sulfur coal producers. For instance, although the NCA purports to represent the entire coal industry, its position actually favors eastern high-sulfur producers. Consequently, the western low-sulfur coal industry has formed its own organization, the Alliance for Clean Energy. Its main goal is to ensure that any legislation including acid rain provisions allows for switching to low-sulfur coal rather than requiring scrubbers. Similarly, western utilities, realizing that EEI represents the interests of the midwestern utility and coal companies, formed a consortium and hired a Washington attorney following a 1984 proposal to extend controls nationwide.[32]

Electric Power Research Institute   EPRI is concerned with the scientific aspects of the debate. Its research also addresses some economic questions and is based on a particular set of assumptions. EPRI is approaching the problem comprehensively, and is supplying much of the data on which other industry-sponsored groups base their positions. However, unlike the National Academy of Sciences and other independent research groups, EPRI urges no immediate implementation of SO_2 and NO_x emissions control measures. While conceding that the electric utility industry is a major source of SO_2 and NO_x, precursors to acidic deposition, EPRI maintains that specific cause-and-effect relationships have yet to be determined.[33]

EPRI's research covers many aspects of acid rain, including pollution emissions, atmospheric processes and deposition, ecological effects, economic assessments, emissions control technologies for SO_2 and NO_x, advanced clean-coal generation technologies, and fuel planning. EPRI is involved in three demonstration projects for fluidized-bed combustion technology, as well as the Cool Water project to demonstrate the integrated gasification combined cycle. Between 1974 and 1986 the utilities industry provided EPRI with $350 million for acid deposition research-and-development activities; $180 million is anticipated for the subsequent three years. Of this, approximately $30 million is destined for environmental effects, $50 million for developing control technologies, and $95 million for alternative clean-coal technologies.[34]

Environmental studies since 1978 have focused on the nature of emissions, how they are incorporated into clouds, how they are transported, what chemical processes occur, and what materials are deposited. Early work in the Adirondack Mountains examined the vulnerability of lakes to acidification. A predictive model (ILWAS) was developed to assess the relationship between the acidity of deposition and that of surface water. Two conclusions were reached: (1) controlling SO_2 emissions alone may not be sufficient to control acidification; and (2) the depth and composition of soil till affects buffering capacity.[35]

EPRI maintains that these conclusions call into question much of the data environmentalists cite estimating the number of lakes endangered by acid rain. Since most acid lakes are found in regions with granite bedrock, many previous investigators had designated large regions of the United States and Canada that have granite bedrock as acid sensitive. However, EPRI examined three lakes located on surficial deposits brought in by glaciers, and found that they had pHs of 7.0, 5.5, and 4.5. The most acidic of the lakes received more surface and shallow subsurface flows from its watershed; the least acidic lake received a large fraction of its inflow from deep seepage. The implication of these findings is that the use of bedrock geology alone to delineate areas sensitive to acid rain is not always sufficient. EPRI researchers believe that "incorporation of some of the results [of these investigations] into the current sensitivity maps would have the effect of reducing the number of lakes that are indeed susceptible to acid precipitation at this time."[36]

At this point in the debate over acid rain, EPRI recognizes a number of "potential problems." Most significant is the acidification of lakes, although EPRI maintains that the relationship to acid rain is yet to be determined with certainty. EPRI's main concern is the lack of quantification in relating emissions to acid deposition. Also significant, EPRI admits, is the documented damage to certain types of plants. Beyond these two problems "there are hypothesized problems of effects on forest productivity, soil acidity, and damage to materials and statuary." EPRI is extending its studies to the factors responsible for forest decline, crop productivity, soil chemistry, and effects on materials. This research is slow to yield results, and much more remains to be done. So far the research has raised more questions than it has answered, the most important of which are: What level of source emissions will cause what precipitation pH values? For a given reduction in source emissions, what will be the change in precipitation pH? How many lakes are nonsupportive of fish due to acid rain? How many lakes will become supportive of fish for a given level of reduction in source emissions? At what level of precipitation acidity will the fertilizing effects of sulfur and nitrogen be outweighed by the harmful effects of hydrogen ions on agricultural crops and forests?

EPRI maintains that any action on the other steps in the process is premature and unwise at this time. "Without answers to these questions, we cannot determine what mitigation strategies may be required; nor can we determine the effectiveness of any remedial measures for acid precipitation effects."[37]

Naturally, environmental groups disagree with many of EPRI's conclusions, and some probably see EPRI's constant warnings about uncertainty as more of industry's informational haze. Unlike other industry groups, however, EPRI has acknowledged the potential seriousness of the problem and is seeking answers to the important questions concerning acid rain. The real source of conflict between environmental groups and EPRI is that, in the face of uncertainty about environmental damage, environmentalists would rather err on the side of protecting the environment, and EPRI would rather err on the side of the economic status quo. If we act prematurely, EPRI argues, we may waste scarce resources on control measures that are unnecessary or ineffective.

Edison Electric Institute   EEI often cites EPRI research and conclusions in support of their policy suggestions, but they always do so to minimize the seriousness of the problem and to emphasize the seriousness of the economic burden that further controls would impose on utility companies and consumers. It is logical that EEI should support utility interests, but their position would seem less suspect to environmentalists and policymakers if their resistance to the environmental position was not so comprehensive. EEI has sought to discount practically every claim of damage to the environment, even at the cost of their own credibility. In House hearings on acid precipitation in 1981, Al Courtney, speaking on behalf of EEI, disputed the causes of acidification, the contribution of power plant emissions, and whether imposing a heavy financial burden was justified in the absence of assured benefits to the public.[38]

All of EEI's questioning of the scientific basis for the fears of acid rain goes ultimately to support the economic argument against the controls proposed by the major control bills. EEI warns against the economic risks of premature regulatory programs, arguing that too many uncertainties exist to establish regulatory requirements. "The expenses associated with a significant emissions reduction will be very great ... [and] simply cannot be justified at the present time in light of the highly speculative nature of the benefits that might result from additional controls."[39]

Much of EEI's argument depends upon inflated cost estimates. A survey of twenty-four utility companies conducted by EEI in the early 1980s concluded that electric bills in some parts of the United States might rise by as much as 50 percent if proposed amendments to the Clean Air Act requiring reductions in sulfur dioxide emissions were enacted.[40] In a more recent analysis of HR 4567, a House bill that would require each state to reduce acid rain pollutants, EEI found that annual costs of implementing the program approached $9 billion, as compared to an OTA estimate of \$3.8– \$4.9 billion.[41] Using EPA data, EEI claims that total SO_2 emissions dropped 28 percent between 1973 and 1985, while coal consumption increased 80 percent. Because changes in rainfall acidity during this time period could not be detected, EEI advocates "continued research, direct mitigation of acidic conditions in lakes, and development and deployment of new clean coal technologies."[42]

National Coal Association   NCA is a traditional lobby representing the interests of coal producers who oppose amendments to the Clean Air Act designed to control acid rain. NCA, like EEI, expends much effort pointing out the scientific uncertainties in the debate, as well as the abuses of scientific material by environmentalists and the press. Unlike EEI, however, NCA has acknowledged the problem as serious, both in current and potential impacts. NCA points out the scientific uncertainties not to suggest that no action need be taken, but that the appropriate action is yet to be determined. Underlying NCA's arguments is the conviction that the coal industry is serving as a scapegoat under the provisions of the current control bills.

Much of NCA's concern revolves around what it perceives to be serious economic consequences to be encountered if further regulations are imposed. Indeed, much of NCA's argument relies upon the results of impact studies by the UMW and the Peabody Coal Company that may overstate the impacts of regulation. A UMW study cited by Bagge estimates that 89,000 miners would lose jobs under one of the pending control bills. However, the UMW study fails to consider that some losses would be offset by increased low-sulfur coal production and by jobs created in pollution control. Further, the study ignores the jobs in outdoor recreation and tourism that will be lost if acid rain is not curbed. Similarly, NCA distributes Peabody Coal Company estimates of increased costs to utilities of \$8.5 billion annually.[43] This estimate is well outside the reasonable limits defined by Congressional Research Service studies.[44]

The ultimate thrust of NCA's arguments, however, is more political than economic or scientific, and NCA has raised several points relating to the fairness of the proposed political solutions. All of the pending control legislation is aimed mainly at coal-burning stationary sources. Research suggests, however, that oil-burning sources produce nitrogen and sulfur compounds that are more readily converted to acids in the atmosphere.[45] Yet coal-burning utilities alone are being targeted to bear the entire burden of control, while oil-burning plants and smelters are exempt from regulation. In addition, NCA has pointed out geographic inequities that, in the past, would have resulted in mid-western coal-producing states bearing a disproportionate share of the cost of control. These concerns have since been addressed, at least to some extent; much of the legislation introduced since 1984 provides for some form of national cost sharing (see chapter 7).

NCA has also been concerned about accusations and "inflammatory rhetoric" from Canada.[46] NCA publicly discounts any Canadian conspiracy to capture a share of the U.S. electricity market by promoting pollution control legislation that would raise U.S. production costs. However, NCA has, in the past, found Canadian accusations and criticism of U.S. pollution control efforts irresponsible, given the shortcomings of Canadian environmental protection and the large amounts expended by the United States in implementing the Clean Air Act.

In recent testimony before the Senate Subcommittee on Environmental Protection, NCA asserted the efficacy of the existing provisions of the Clean Air Act and argued that no additional measures were necessary. In NCA's view, additional controls would cost a great deal of money and yield marginal benefits. Four reasons were given for the success of the current program in reducing SO_2 emissions, and for why continuing reductions should be expected: (1) new-source performance standards; (2) the shift in demand to regions with low-sulfur coal; (3) substitution of lower-sulfur coal in power plants throughout the country; and (4) addition of scrubbers to plants not subject to new-source performance standards.[47]

American Electric Power Service Company   AEP of Columbus, Ohio, generates more power than any other electric utility company and has been regarded as an efficient operator of its coal-fired plant network. AEP consists of eight electric public-utility operating companies serving customers in approximately three thousand communities in seven states in the east-central and midwestern United States. While other electric companies in the northern United States and Canada were hurt by the slowdown in the growth of electricity use over the last decade, AEP sales have grown, mostly through wholesale sales to other utilities. Neighboring utilities often buy peak-period power from AEP more cheaply than they are able to produce the power themselves.

In 1982, however, AEP's overall electricity sales were off 12 percent and sales to other utilities dropped more than 20 percent. Also, AEP has a 25.5 percent interest in a troubled nuclear plant that is under construction. These reasons make new expenditures on emissions controls difficult for AEP to bear. AEP spokesmen state that the Midwest offers few attractions to industry over other areas of the country. One advantage, however, is relatively low electricity rates, which pending clean air legislation could wipe out. The loss of this advantage could cause some industries to relocate. AEP believes that not enough is known about the effects of acid rain to justify such high costs and is against any kind of amendment to the Clean Air Act that would mandate reductions in SO_2 emissions.

AEP has circulated a report entitled "Economic Impact on the AEP System of Compliance with the [1981] Mitchell Bill." The analysis purports to assess the measures that would be needed to achieve the required ten-million-ton SO_2 emission reductions, the annual cost of such measures, and the extent to which those costs would increase the electric bills of AEP's consumers. The estimates given in this report are almost an order of magnitude higher than estimates in other industry and nonindustry analyses and, when reflected in electrical rates, result in a startling 63.3 percent increase.[48] The estimates have been published in newspaper and magazine ads, distributed on pamphlets with utility bills, and circulated to congressmen involved with reauthorization of the Clean Air Act.

The AEP estimates are based on several misconceptions about the proposed amendments to the Clean Air Act. Nearly all other cost estimates indicate that rate changes would not approach the AEP estimates.[49] Even if utility bills increased as much as AEP estimates, the rates would still be lower than those paid in the areas of the country hardest hit by acid rain. With recent bills (such as HR 4567) limiting local residential rate increases to 10 percent, it is going to be more difficult for AEP to make a case against controls. However, this fact has not stopped a lobbying organization, Citizens for the Sensible Control of Acid Rain, from claiming that acid rain controls will result in 30 percent rate increases. This group, founded in 1983 to fight acid rain controls, is supported by utility companies, who contributed nearly $3 million between September 1985 and June 1986. American Electric Power topped the list of contributors at $1.5 million.[50]

Tennessee Valley Authority TVA is a public enterprise. Its role as utility provider and guardian of resources for the Tennessee valley makes it an interesting comparison with utility interest groups. Traditionally TVA has been aligned with other utility interests on regulatory matters. On the acid rain issue, however, TVA's position departs from that of the privately owned utilities on most major points. Several reasons for this departure are possible; one important factor was the presence of S. David Freeman, a recognized expert in energy and conservation issues, on the TVA board. Freeman, who left TVA in 1984, has made the board more sensitive to its role as guardian of resources.[51]

TVA's position is stated in testimony submitted to congressional committees and in position papers circulated by TVA.[52] Its departure from the position of the utility interests on the scientific aspects of the debate is evident from a TVA position paper: "TVA recognizes the likelihood of a relationship between acid precipitation and the total load of SO_2 and NO_x in the atmosphere. TVA also recognizes that long-range transport and transformation of pollutants in the plumes of fossil-fueled boilers of all types produce sulphate and nitrate particles which are believed to be linked to acid precipitation. Moreover, acid precipitation is not only a local problem, but a regional, national, and international problem that cannot be adequately addressed by present requirements under the Clean Air Act."[53]

TVA's position on the ethical argument is implicit in the broader understanding of regulatory costs and benefits that it supports. In another position paper, discussing the impact of two particular control bills, TVA states: "These costs [of complying] can be estimated reliably within a range that reflects varying operating assumptions. Benefits are more difficult to measure but nonetheless significant. This is especially the case if long-term and regional benefits are included in addition to more easily quantified local benefits. TVA believes that any discussion of pending acid rain legislation should take into account these benefits as well as costs."[54]

This broader understanding of long-term, widely ranging (and not necessarily quantifiable) costs and benefits has been lacking in most utility industry statements. TVA estimates that costs to rate payers could increase by 3–13 percent as a result of acid rain control amendments, depending on the particular bill and on varying operating assumptions.[55] This range stands in considerable contrast to AEP's estimate of increases up to 66 percent, even though both TVA and AEP have about the same coal-fired capacity.[56]

TVA is also arguing the political issue of the fairness of the kind of regulation that is proposed in pending acid rain bills:

TVA believes that any bill that attempts to reduce emissions across the board is unfair by tending to discount control measures already in place. More specifically, TVA believes that legislation must recognize that a major contribution to the total load of SO_2 is produced by generally older and smaller stationary sources currently operating at high levels of SO_2 emissions from untreated high-sulphur coals. TVA believes that a workable legislative approach to reduction of these pollutants is best found by capping the pollutants at all stationary sources. A cap on individual source emissions in the range of 4 to 5 pounds of SO_2 per million BTUs represents a reasonable level of control.[57]

Although there does not seem to be any movement by policymakers to accommodate TVA's argument, one could argue that TVA has increased its credibility and its chances for influencing the specifics of regulation by conceding key points on the other aspects of the debate.

The Center for Clean Air Policy

The Center for Clean Air Policy, created by a number of state governors, hosted a series of analytical studies over a sixteen-month period directed at two distinct aspects of the acid rain debate: (1) the costs of various control options for two major utilities, AEP and TVA; and (2) a dialogue between the interested parties in order to identify areas where agreement could be reached. The following discussion is taken from the ensuing publication, *Acid Rain: Road to a Middleground Solution.*[58]

In preparing cost estimates, three main policy options were chosen for analysis. Each was analyzed under various implementation strategies by ICF, Inc. AEP and TVA together accounted for almost 10 percent of the total electricity generated and about 17 percent of the SO_2 emissions in 1985. Cost estimates to each entity were generated by ICF, and the assumptions used in deriving these estimates were compared with those used by the utilities. The data presented for AEP illustrate the problems associated with projecting the cost of implementing a nationwide emissions control program. Annualized cost estimates to AEP associated with SO_2 reductions of 0.9–1.1 million tons (under a national reduction scenario of a ten-million-ton reduction) were estimated by ICF to be $0.2–$0.5 billion. This would result in average electricity rate increases of 3–9 percent and is considerably less than AEP's least-cost estimate of $0.9 billion. The main areas of disagreement in these two estimates derived from differences in assumptions concerning: the changing rate of utilization of various plants; annualization rates for control costs (over 20–30 years rather than 10 years); and projections for low-sulfur coal premiums (a factor of two, resulting in less reliance on scrubbers in the ICF scenarios).

The dialogue process involved representatives from thirty diverse organizations, including environmentalists, utilities, states, and consumer groups. The goal was to define where the group could agree and where areas of major disagreement persisted. Because most participants had to settle for options that they considered less than optimal (but not totally unacceptable), the effort usually resulted in the selection of each participant's second or third choice. In this way broad areas of agreement could be discerned. These included: the need to encompass a wide range of environmental benefits; a reduction target of 8—10 million tons of SO_2 emissions; the inclusion of all sources emitting more than 100 tons of SO_2 per year; maximum flexibility in selecting the control option; and tonnage reduction requirements as opposed to average emission rate reductions.

Disagreement persisted over the question of whether the regional economic impacts of controls were sufficient to justify subsidies. Representatives from the lower Midwest were willing to forgo this concession if controls were based on reduction allocations per state and not targeted specifically at utility emissions. However, no consensus was forthcoming on how a suitable formula could be derived.

Canada: Policy from the Outside In

The major Canadian environmental and industrial groups substantially agree on the scientific and ethical arguments about acid rain. Both concur that there is enough scientific evidence to conclude that acid rain is a danger to the environment and that man-made emissions are a major contributor to the problem. Both groups now take the position that, in the face of uncertainty about environmental effects of emissions, it is better to err on the side of the environment than on the side of short-term economic benefit. Although the costs of abatement may be great, the two groups agree that U.S. and Canadian industry should begin to control emissions either by switching to alternative fuel sources, by conservation measures, or by equipping existing and new coal-burning plants with emissions control equipment. The major Canadian environmental and industrial groups also agree that some cooperative action must be taken soon by the two governments.

Canadian environmental and industrial groups recognize airborne pollutants as a major cause of acid rain and recommend reducing emissions in order to protect the fishing, tourism, and forestry industries in the eastern provinces and in the northeastern United States. Both groups acknowledge that Canada does contribute to the acid rain problem, and the major Canadian industries involved in the problem have initiated steps to limit SO_2 emissions, the presumed major source of acid rain. It is also recognized that acid rain is harmful (and primarily man-made), that SO_2 emissions can be transported long distances, and that acid rain is lowering pH levels in lakes and rivers in eastern North America to the point where they can no longer support fish life.

Canadian interest began in 1971, when R. J. Beamish and Harold Harvey published their report entitled "Acidification of the La Cloche Mountain Lakes," which documented the decrease in fish life in lakes in the Killarney Park wilderness region in northern Ontario.[59] Beamish and Harvey had studied the region since 1966, when they tried stocking the lakes in order to reestablish the level of sport fishing for which the area had once been acclaimed. When none of the four thousand pink salmon that Harvey and Beamish introduced survived the first year, they decided to measure the acidity of the area's lakes.

During the summers of 1969−71 Beamish and Harvey tested sixty lakes in the region and discovered that the pH level averaged 4.4, a dramatic drop from the 6.8 pH measured ten years earlier. The ten-year pH change in the Killarney lakes represented a more than hundredfold increase in acidity. Although Harvey's report concentrated on his observations of declining fish life in the Killarney lakes, he also referred to the 1971 Swedish Royal Ministry for Foreign Affairs and Royal Ministry of Agriculture report linking the acidification of Swedish lakes with pollution from industrial sources upwind in England and other parts of Europe. According to the Swedish study, pollutants could remain airborne much longer and thus travel much farther than had previously been believed. Since the Killarney lakes are located downwind from the International Nickel Company's (INCO) smelter at Sudbury, Ontario, Harvey suspected that airborne pollutants from the smelter could be responsible for the increased acidity of, and consequent declining fish life in the Killarney lakes. Thus Harvey introduced the term "acid rain" into North American terminology.

Despite Harvey's predictions of potentially dangerous damage to the Canadian environment, few Canadians were roused into action. Not until 1976, when Ross Howard, a reporter for the *Toronto Star*, wrote articles on acid rain based on conversations with Harvey did public reaction force government officials and industry leaders to rethink environmental policies.[60]

Canadian Environmentalists

Canadian Coalition on Acid Rain   CCAR represents more than fifty groups, mostly from eastern Canada.[61] Founded in June 1981, CCAR does not advertise itself as an environmental group. (Less than one-third of the member groups are primarily environmental groups.) It has no official (but a close informal) link with NCAC in the United States, and has never officially supported any specific amendments to the Clean Air Act. Funding for its Canadian activities is provided by the Canadian government, although it is not officially linked with the government. CCAR is primarily interested in informing the public and government and business leaders on both sides of the border of the dangers of acid rain. By remaining independent and disseminating information, CCAR believes it remains credible and convincing.

CCAR applauds the attempts by INCO and Ontario Hydro to limit SO_2 emissions. In 1983 Michael Perley, co-director of CCAR, stated that: "INCO and Ontario Hydro have recognized the correlation between SO_2 emissions and acidic lakes, and have aggressively pursued policies to reduce SO_2 beyond the nonappealable orders issued by the provincial government of Ontario."[62] Perley was confident that both of the major contributors to the acid rain problem on the Canadian side of the border are serious in reaching reduction goals. However, CCAR doubts that either company would have agreed to the reductions mandated by the provincial government of Ontario if the three major political parties were not in full agreement on the subject. With no political backing, INCO and Ontario Hydro had little choice.[63]

According to CCAR, Ontario Hydro is not spending any money on abatement equipment because the company is transferring from coal- to nuclear-powered generating plants.[64] The transfer can be made easily for three reasons. First, Ontario Hydro has been planning to convert to nuclear power since the first shock of oil price hikes, and several new plants are being brought on-line. Second, the company does not have to deal with the encumbering and expensive regulations adhered to in the United States. Because Canada does not guarantee freedom of information, does not require environmental impact statements, and does not recognize citizens as having any standing in court on environmental matters, Ontario Hydro can choose cost-effective methods with few administrative obstacles. Finally, Ontario Hydro is a regulated company of the Crown and can pass any price hike due to construction costs on to consumers.

CCAR maintains that officials from INCO have been considerate and modest about the company's attempts to reduce SO_2. Nevertheless, CCAR was concerned about what would happen when production at the Sudbury plant was resumed after an upturn in the nickel market. INCO has installed abatement measures at the plant, but CCAR believes that Canadian citizens will pay for them in several ways. First, INCO will write off about 50 percent of the cost of abatement equipment; second, INCO will write off another 20–25 percent on energy efficiency; and third, the remaining costs will come in the form of a handout from the Canadian government.[65]

CCAR firmly believes, based on the studies available, that SO_2 and NO_x emissions are the main cause of acid rain. They insist that some controls are immediately necessary on both sides of the border and they are working to inform anyone who will listen. They believe that the states presently not affected by acid rain (the western and deep southern states) are the key to a conflict resolution, and they are trying to educate leaders and groups from those areas about the evidence gathered by many concerned scientists.

Canadian Industry

The Canadian firms discussed here, Ontario Hydro Electric Company and INCO, Ltd., were selected for three reasons. First, both are located in eastern Canada, where the major impacts are felt. Second, both are among the largest public and private emitters of SO_2 in eastern Canada. In fact, Ontario Hydro is the major utility in eastern Canada that burns a significant amount of coal for energy. Finally, both are major examples of the cooperation between Canadian industry and the government in reducing SO_2 emissions.

INCO, Ltd.   The twelfth-largest company in Canada, with $4 billion in assets, INCO, Ltd. (formerly International Nickel Company) has dominated the world nickel markets since 1902. INCO once produced over 90 percent of the world's supply of nickel and remains the world's largest single producer. Headquartered in Toronto and New York, INCO owns mines and smelters throughout the world, including the world's largest smelter in Sudbury, Ontario.

At one time INCO's Sudbury smelter was the most photographed and discussed source of SO_2 emissions in North America. Throughout the sixties more than 6,000 tons of SO_2 poured out of the stacks at Sudbury each day—the largest source on the continent. By 1970 the yellow-brown skies over Sudbury were causing recurring violations of ambient air standards, prompting the provincial government of Ontario to press for reforms. By 1973 INCO had reduced emissions to 3,600 tons per day, mostly by constructing a mill to extract sulfur from the nickel ore before smelting. No further steps were taken until 1978, when a joint U.S.-Canadian report triggered a flood of attention and publicity about acid rain. Realizing that INCO could not achieve the previously ordered reductions (to 750 tons per day by 1979), the government of Ontario imposed a nonappealable regulation to reduce emissions to 2,500 tons per day beginning in September 1980. The regulation required INCO to further reduce emissions to 1,950 tons per day by January 1, 1983.[66]

During the final months of 1982 and much of 1983, INCO's smelter in Sudbury was closed as a result of depressed world nickel prices. Following ten consecutive quarters of losses, it resumed production in late 1983. Environmentalists were concerned that INCO would not meet the proposed standards when the Sudbury plant reopened, but when it returned to full production INCO's emissions of SO_2 averaged about 1,900 short tons per day. This reduction was achieved mainly through relatively cheap ($14 million) milling modifications that reduce the sulfur content of nickel before smelting.[67] INCO does not consider environmental regulation as a significant contributor to its economic woes, but continued economic losses could damage efforts to implement emissions-reducing technology.

New, stricter SO_2 reduction targets imposed in 1985 require a number of technological improvements in order to reduce the amount of sulfur entering the furnace and increase the amount of sulfur captured from gaseous emissions. INCO reports that it has spent more than $100 million on research and development in this decade.[68] However, if these newer, more efficient technologies are not available to meet reduction target deadlines, depressed nickel prices may make it necessary for the government to subsidize control costs. The federal government has allocated $150 million for emissions controls for the Canadian smelting industry.

U.S. utilities and midwestern politicians claim that, while INCO's proposal to install a process that "has the potential of converting as much as 80 percent of the sulfur in the nickel concentrate into a continuous stream of sulfur dioxide suitable for conversion into sulfuric acid"[69] appears promising, the costs involved are too great. They believe that INCO's only reason for announcing the plan was to deflect criticism toward coal-fired electric generating plants in the United States.[70] Furthermore, skeptics believe that the process will never be implemented and that it may be an attempt to sway the U.S. public to force legislative mandates for reductions in SO_2 and NO_x emissions. INCO officials maintain that they favor the incorporation of emissions reduction targets for SO_2 and NO_x as a part of the reauthorization of the Clean Air Act, and counter criticism by saying that they, at least, admit they are a part of the problem and are proposing measures to remedy the situation. INCO supports the March 1984 decision of the Canadian government to reduce emissions unilaterally below 1980 allowable levels by 25 percent by 1990 and by 50 percent by 1994.

Ontario Hydro Electric Company   The first emissions reductions imposed on Ontario Hydro in 1981 required a 50 percent reduction by 1990 (as discussed in chapter 6, more stringent reductions were imposed in 1985). As a result, Ontario Hydro chairman Hugh Macauley announced a greater reliance on nuclear power, effective conservation programs, and expansion of hydroelectric power. He maintained that these measures constitute a better environmental and economic alternative to a policy of installing scrubbers on an increased number of coal-fired plants, which contribute only about 3–5 percent of the acid rain falling on sensitive areas in Ontario.

Ontario Hydro is prepared to add scrubbers, if necessary, to existing coal-fired plants that are currently washing and burning low-sulfur coal. But coal is the swing fuel of Ontario Hydro's system and is generally used only to meet peak and intermediate loads.[71] Because the coal-fired plants are used less and less as a result of decreased demand, and because twelve new nuclear plants will be coming on-line during the coming decades, Ontario Hydro may be able to shut down existing coal-fired plants totally and thus eliminate SO_2 and NO_x emissions. From Ontario Hydro's perspective, the best available technology for reducing acid rain is the Candu nuclear reactor.

Ontario Hydro has been criticized for backing away from early commitments to employ scrubbers on coal-fired plants. Critics charge that Ontario Hydro has failed to set a good example (by installing scrubbers), and has intensified the reluctance of American utilities to initiate any control strategy. The American utilities can point out that, while Ontario Hydro may advocate control strategies, the company has not taken any real or costly steps toward implementing one. Some critics believe that Ontario Hydro's position on scrubbers is designed to place pressure on American utilities to adopt costly control strategies, thus allowing Ontario Hydro to undersell the American utilities in eastern U.S. markets.

This "Canadian conspiracy" speaks more for the ability of Canadian utilities to shift from coal-fired plants to nuclear and hydroelectric plants than for their greater concern for the environment. A report by the Congressional Research Service states: "More importantly, if U.S. utilities look to Canada for nuclear power, it is more of a commentary on the state of the U.S. nuclear industry than on acid rain legislation. Acid rain legislation provides American nuclear power with the same incentives as Canadian nuclear power. If U.S. utilities decide to get nuclear power from future Canadian plants, it will be due to their assessment that the nuclear option in the United States is not an option for them, and therefore the regulatory, technical, and political difficulties in obtaining power from Canada are worth the risks."[72]

But even if Canada possesses excess electricity for sale in the United States, regulatory, technical, and political difficulties inside Canada would severely hinder any export sales. From a regulatory standpoint, Canada avoids long-term, firm export contracts in order to ensure sufficient electricity for its own demand.[73] Export contracts are generally short-term and seasonal. (Canadian domestic demand peaks in winter, while U.S. demand usually peaks in summer.) There are two situations, however, under which electricity for export would be available. The first possibility would be if the Canadian government projections for domestic electricity demand are significantly higher than actual demand. The second would occur if the building of Canadian nuclear power plants is partially dedicated to export.

The first scenario has a logical basis: If Canada were to have excess electrical power, it could be sold in U.S. markets. However, this would require that new cables be built across the border, creating more regulatory and international problems than the project may be worth. The second scenario would have serious political implications, chief among them being the likelihood that the Canadian people may wonder why Canada should assume the risks of nuclear power plants while the United States consumes the electricity.

The Alignment of Canadian Interests

By favoring the acid rain amendments under consideration in the U.S. Congress, INCO, Ontario Hydro, and the Canadian Coalition on Acid Rain have allied themselves with the eastern provincial and federal governments in Canada, affected states in the northeastern United States, and the National Clean Air Coalition. This undefined and unattached consortium believes that overwhelming but circumstantial scientific evidence indicates a relationship between the amount of sulfur and nitrogen oxides put into the air and the amount of acid found in lakes hundreds of miles away. These groups call for immediate action that would mandate the control of emissions from coal-fired utility plants in the eastern United States. They believe that the utility-related industries in the United States confuse the public and policymakers with cost-benefit analyses that pay lip service to environmental and social costs.

The role of Canadian interest groups in U.S. policy formulation is necessarily problematic. Although the Canadian government and interest groups should have avenues available for expressing their opinions on U.S. policy matters that directly affect Canada, Canadian interests probably do well to avoid promoting specific political measures. The role of public educator that CCAR has chosen is probably the most likely to achieve the desired goal. One of the ironies of the acid rain problem is that Canada may be in a better position to control emissions because of weaker Canadian environmental laws, which have facilitated the shift to nuclear power.

Conclusions

Many observers have concluded that the debate over acid rain has now shifted. According to the National Wildlife Federation: "A public increasingly concerned—and informed—about acid rain has shifted debate on the issue on Capitol Hill. The big question is no longer whether to enact control legislation, but rather how much control is needed . . . and how to allocate costs."[74]

Nor is this sentiment just propaganda by the NWF. In 1983 William McGonnell of EEI also agreed that something similar to the Mitchell amendment to the Clean Air Act would eventually pass. But, he added, "It will be a big mistake."[75] Similarly, Larry Parker of the Congressional Research Service suggested that proponents of controls were mistaken in scoffing at the economic issues raised by utilities. Parker implied that the real argument at this point is an economic one and that once the economics of regulation were satisfactorily provided for, a policy could be established.[76] While there is still much research to be done and many political questions to be answered, the delay in acid rain policy formulation is caused by economic considerations.

The position taken by environmentalists on acid rain has benefited from increasing public interest over the last few years. One reason is that this position is simpler for the public to understand, and it evokes an immediate emotional response. In addition, environmental groups are better organized and more politically astute than in the past, and they now conduct sophisticated media campaigns. Perhaps more important, the public is reacting to the political atmosphere in the country, in particular to environmental policies and appointments of the Reagan administration. The disarray at EPA under Anne Burford Gorsuch and the highly publicized activities of James Watt alarmed the public and contributed to the perception that Reagan is antienvironment. Sensing that these policies are resulting in a perception that the Republican party is antienvironment, a group of Republican members seeking control of the House by 1992 agreed to work with Representative Henry Waxman in the spring of 1986 to develop a compromise bill containing acid rain control provisions.[77]

Industry groups have contributed to this public reaction by opposing environmental regulation on all fronts, even in areas where they
lack credibility. The application of cost-benefit techniques to environmental matters has caused considerable discontent. Much of the public, and all environmental groups, view these techniques as inappropriate and shortsighted. In addition, many industry attempts at
cost-benefit analysis have been demonstrably inaccurate as well. By
persisting in opposing all of the claims made by environmentalists,
and by using techniques that are of doubtful value in environmental
matters, industry groups have damaged their credibility, even on
the questions to which they can make a legitimate contribution. Thus,
industry is likely to suffer from policies that will be formulated in
spite of their arguments, rather than in consideration of their
arguments.

The Canadian government and interest groups have certainly
enhanced public awareness of the acid rain problem. However, Canadian interests have not always been pursued as effectively as they
might be. CCAR has settled into a role as public educator that empha-
sizes the mutual interest of the United States and Canada in controlling acid rain. This role will probably be an effective and legitimate
one. However, many of the scathing political comments by Canadian
government officials in the past probably reinforced the recalcitrance
of U.S. officials to cooperate in regulatory efforts. In addition, Canada
was an easy target for criticism on environmental matters because
regulatory measures there were generally less stringent than in the
United States and because Canada offers few opportunities to environmental groups to influence policy-making. Certainly a part of the
Canadian position must be its continuing efforts to impose stricter
regulations on Canadian industry.

The suggestion that Canada should clean its own yard first has been
criticized as the "shining example theory of diplomacy" by Don
Munton.[78] While Munton is probably correct in assuming that Canadian cleanup efforts will not necessarily inspire the United States to
do the same, the Canadian cleanup will certainly help to remove some
major barriers to negotiation and policy formulation. The Reagan
administration, which is not hospitable to the idea of further regulation, will certainly take advantage of any Canadian hesitation to act. In
addition, weak environmental regulation in Canada has afforded utility interests a tool to influence legislators. Clearly, a part of the political solution is for Canada to accept a large share of responsibility for
the cleanup.

Chapter Six
Canada's Acid Raid Policy:
Federal and Provincial Roles

||||||||

Introduction: The Canadian View of Acid Rain

Because acid rain is most prevalent over the northeastern part of the North American continent, it has the greatest impact on Canada, affecting some of its most populous regions. As a result, acid rain is considered to be a much more pressing problem in Canada than in the United States. Estimates show that 77 percent of the Canadian public view acid rain as their most serious and pressing environmental problem.[1] In addition, meteorological patterns in North America are such that the United States is generally believed to be a net exporter of sulfur and nitrogen pollutants to Canada. The Canadian government estimates that: (1) roughly one-half of the sulfur in Canadian acid rain originates in the United States; and (2) three to four times as much sulfur travels north across the border as goes south to the United States.[2]

The Canadian provincial governments believe that acid rain has caused extensive damage to aquatic and terrestrial ecosystems in eastern Canada. They estimate that 140 lakes in Ontario alone are without fish as a result of acid rain, and that thousands of lakes in Ontario, Quebec, and Nova Scotia are "acid stressed" and in danger of destruction.[3] The Ontario government predicts that damage to its lakes may result in a loss of six thousand jobs and as much as $28 million in annual revenue from tourism and fisheries.[4] Other cost estimates include $1 billion in damage to Canadian buildings every year, and damage to the timber industry, which accounts for $20 billion in

This chapter was written by Robert Egel.

annual revenues and directly or indirectly provides one million jobs.[5] The *New York Times*, noting that acid rain in Canada affects the most populated parts of the country, estimated that 8 percent of the Canadian gross national product is believed to be at risk from possible damage to lakes, forests, and croplands.[6]

Canada produces electric power from a diversity of sources, including nuclear, but the major sources of SO_2 in Canada are nonferrous ore smelters (44 percent of SO_2 emissions). In the United States thermal power plants are responsible for 65 percent of SO_2 emissions, contributing to a total of 28.5 million tons. Total Canadian emissions of SO_2 are 5.5 million tons per year. The major sources of NO_x in both countries are automobiles and industry. The United States produces 24.4 million tons per year, more than ten times as much as Canada.[7]

As a result of these factors, the nature of the debate about acid rain differs both qualitatively and quantitatively in the two countries. Because acid rain is perceived to have a much greater impact on the economy as a whole, and because Canada sees itself as the victim of U.S. pollution, there is much greater agreement there about the need to control emissions. In discussing the Canadian approach to reducing acid deposition, this chapter focuses on the domestic policies pursued in Canada and the options available to the federal and provincial governments for their implementation. Specifically, it will discuss: (1) the existing legal basis for the federal government to control emissions that cause acid rain; (2) the policies that have been adopted by the federal government to do so; and (3) the legal basis and policies of two Canadian provinces—Ontario and Quebec—for controlling emissions of acid rain precursors.

Federal Authority to Control Acid Rain

The approaches taken by Canada to control air pollution, and specifically acid rain, differ in several significant respects from the options available to the United States. This stems from differences in the constitutional role of the federal governments and from a far less litigious approach to conflict resolution in Canada. The Canadian federal government is primarily concerned with interprovincial and international issues. Because its constitutional role in domestic legislation is not well defined, most domestic issues are settled through negotiation with the provinces early in the process, rather than in the courts after the enactment of legislation. However, because it operates under a parliamentary system, the government speaks with a more unified voice and is more effective in negotiating international agreements.

In the case of air pollution control, the responsibility for control of stationary (industrial) sources of air pollution rests almost entirely with the provincial governments. The federal government has exclusive authority over interprovincial and international transboundary pollution control. For mobile sources the responsibility is shared: the federal government defines standards for vehicle manufacture, and the provinces regulate and inspect vehicles in use.

The Clean Air Act of 1971

The Clean Air Act[8] is the primary legislative mechanism for controlling air pollution in Canada. For effective implementation the act is dependent upon close cooperation between the federal and provincial governments. Under the Clean Air Act the minister of the environment is empowered to establish a national system of air quality monitoring stations, commission air pollution research, formulate comprehensive plans for air pollution control, and publish and distribute information to the public.[9] Moreover, the Clean Air Act authorizes the minister of the environment to propose National Ambient Air Quality Objectives, National Emission Guidelines, National Emission Standards, and Specific Emission Standards. Compared to the U.S. Clean Air Act, the federal government's powers are limited and the provinces remain the principal actors.

National Ambient Air Quality Objectives   Section 4 of the Clean Air Act provides that the minister of the environment may formulate National Ambient Air Quality Objectives reflecting three ranges of air quality for each contaminant: "tolerable," "acceptable," and "desirable."[10]

The *tolerable* range indicates a level where there is a danger to the public health; this is roughly equivalent to the primary air quality standard of the U.S. Clean Air Act. The *acceptable* range indicates a level where "welfare" effects may occur, such as damage to the vegetation, water, soil, or to the general public's comfort; this is comparable to the secondary standard of the U.S. Clean Air Act. The *desirable* range indicates a long-range air pollution control goal.[11]

These objectives are only goals and have no legal effect unless they are incorporated into provincial environmental laws. It is therefore up to the provinces to enforce the objectives.[12] Air quality objectives for SO_2 (and other air pollutants, including NO_x) have been published by the federal government for the guidance of the provinces. Most provinces base their control programs on the acceptable level; two use the desirable level.[13]

There are several difficulties with using the National Ambient Air Quality Objectives to control sulfur and nitrogen emissions. First, the provincial governments are not required to adopt them. Consequently, these objectives will be an effective regulatory tool only as long as the provinces choose to act. The Canadian House Subcommittee on Acid Rain pointed out this problem and commented: "There are times, however, when it is necessary for there to be centrally-enforced environmental law standards on a Canada-wide basis—the far reaching consequences of continued interprovincial unabated acid rain is one of these occasions."[14] The subcommittee went on to urge the federal government to develop comprehensive guidelines but did not recommend a shift of final authority to the federal level.[15]

National Emission Guidelines   Section 8 of the Clean Air Act authorizes the federal government to establish National Emission Guidelines indicating maximum "quantities and concentrations [above] which any air contaminant should not be emitted into the ambient air from sources of any class, whether stationary or otherwise."[16] National Emission Guidelines are source specific. They have traditionally been developed by industry-government task forces, which have based the guidelines on the "best practical technology" for control of emissions, rather than on health effects.[17] Thus far, guidelines have been developed for seven industrial categories, including thermal power plants. In April 1981 the minister of the environment announced National Emission Guidelines for new coal-fired thermal power plants. The guidelines were based on commercially available technologies that could reduce emissions of nitrogen oxides by 50 percent and sulfur dioxide emissions by 90 percent, although they did not recommend using any particular technology.[18]

In order to be legally enforceable, National Emission Guidelines, like National Ambient Air Quality Objectives, must be adopted into provincial environmental laws.[19] This makes it difficult to use National Emission Guidelines to reduce nitrogen and sulfur emissions. Even if these guidelines were adopted by all the provinces, other difficulties would remain. The guidelines for coal-fired thermal power plants apply only to new sources, not to old ones, and consequently one would not expect to see a significant reduction in SO_2 or NO_x emissions until most of the old power plants have been replaced by new ones, a process that could take several decades. The House of Commons Subcommittee on Acid Rain therefore recommended: "The

Subcommittee believes . . . that these Guidelines should apply to all coal-fired thermal power plants whether they be new, already-existing, or those which have been converted from oil and gas. . . . Acid rain will not wait for excessive caution—its effects are too insidiously devastating."[20] Finally, the guidelines are pegged to the maximum concentration of contaminants in the ambient air. Consequently, as with the National Ambient Air Quality Objectives, the total amount of the pollutant emitted, which is the more important measure for reducing acid rain, is ignored.

National Emission Standards   Under Section 7 of the Clean Air Act, the federal government is given authority to prescribe compulsory National Emission Standards. These establish maximum concentrations of air contaminants that may be emitted into the ambient air from any class of stationary sources. But the authority to set standards is limited to situations where emissions present a significant threat to the public health or where Canada would otherwise violate an international air pollution agreement.[21] These standards do not need to be adopted by the provinces and can be enforced directly by the federal government. So far, however, this authority has been used only in a few cases involving health effects of toxic substances such as mercury, vinyl chloride, and lead.[22]

No standards have yet been developed for SO_2 or NO_x under this provision, since neither is known to have serious public health consequences, nor has an international agreement been entered to reduce the emissions of these substances. Even if an international agreement were made giving the federal government direct control of SO_2 and NO_x emissions, such an extension of federal powers would probably be met with considerable opposition from the provinces. A more likely approach would be for the federal government to seek provincial agreement in implementing any international obligations.[23] However, given recent reduction targets of 50 percent agreed upon with the provinces (see the next section), it is unlikely that further action on the Canadian side would be necessary should an agreement be reached with the United States.

Specific Emission Standards   Under Sections 20 and 21 of the Clean Air Act, the federal government is given the authority to enact Specific Emission Standards that establish maximum concentrations of air contaminants from stationary sources within any province that has adopted National Ambient Air Quality Objectives. The Specific Emission Standards are to correspond to the maximum "acceptable" level of the objectives, and they must take into account such factors as best available technology, rate of emissions, and total quantity of the pollutant emitted within the whole province.[24] These standards can be directly enforced by the federal government, and provincial approval is not necessary.

Alberta, Ontario, Saskatchewan, Manitoba, Quebec, and New Brunswick have adopted National Ambient Air Quality Objectives for SO_2. The federal government, therefore, could issue Specific Emission Standards for significant SO_2 emitters in these provinces. As with the National Emission Standards, the extension of federal authority in issuing a Specific Emission Standard is likely to be met with opposition, and the federal government would be reluctant to use such power. A more politically acceptable approach would be to secure provincial approval.[25]

December 1980 Amendment to the Clean Air Act   In December 1980 Parliament amended the Clean Air Act to allow the minister of the environment to issue Specific Emission Standards to any source of air pollution if the emissions from that source constitute a significant danger to the health, safety, or welfare of persons in another country, and if that country gives the same rights to Canada in dealing with the effects of transboundary pollution.[26] If the source of air pollution is not a federal project, the minister of the environment must first consult with the province where the pollution source is located to determine if the emissions can be controlled effectively by provincial laws.[27]

The major purpose of this amendment was to establish the authority needed to qualify for the reciprocity requirement of Section 115 of the U.S. Clean Air Act.[28] Section 115 states that if emissions from a source are polluting an area outside the United States, all relevant state implementation plans[29] should be altered to reduce emissions, provided that the foreign country offers U.S. citizens affected by foreign pollution similar protection. To implement the controls required, the minister of the environment could promulgate a Specific Emission Standard. However, the minister would be reluctant to use such a politically sensitive tool.

The Motor Vehicles Safety Act

It was estimated that in 1980 the transportation sector contributed one million tons of NO_x, 50 percent of the NO_x emissions in Canada.[30] The act established a standard for NO_x emissions for new automobiles at 3.1 grams per mile, roughly three times higher than U.S. standards for 1981 automobiles. The Canadian Department of the Environment announced in May 1982 that a socioeconomic impact analysis would be conducted to determine the feasibility of reducing NO_x emissions to 1 or 2 grams per mile and, if recommended, such a reduction would be promulgated. These standards would apply only to new cars because monitoring and regulating cars in use is within the jurisdiction of the provinces. Thus, even by setting the standard at 1 gram per mile, it would take many years before a significant reduction in automobile NO_x emissions would be seen.

Nevertheless, in March 1985 the Mulroney government followed through on a campaign promise and announced that automobile emissions standards for hydrocarbons, nitrogen oxides, and carbon monoxide would be strengthened in 1988-model cars to conform to U.S. standards.[31] These standards are expected to result in a reduction in automobile emissions by more than 45 percent by the year 2000.

Summary

The federal government has only a limited legal basis for the regulation of sulfur dioxide and nitrogen oxides to control acid rain. The Clean Air Act was developed to reduce *local* air pollution, and none of its provisions directly addresses the reduction of *overall* loading of sulfur and nitrogen pollution into the atmosphere. Overall reduction, however, is essential in solving the acid rain problem. In addition, the requirement that both the National Ambient Air Quality Objectives and the National Emission Guidelines must be adopted by the provinces before they can be enforced reduces their effectiveness as regulatory tools.

Moreover, the federal government is unlikely to use some statutory options available to it because of the political liabilities involved. National Emission Standards, which are enforceable by the federal government, could be used under an international air pollution agreement to reduce SO_2 and NO_x emissions. But the federal government would be reluctant to invoke its powers and would more likely depend on the provincial governments to enforce such an agreement.

Specific Emission Standards, which are also directly enforceable by the federal government, could reduce emissions by major SO_2 and NO_x polluters in some provinces. But because use of this provision would likely be met with provincial opposition, the federal government would once again hesitate to use its authority.

The 1980 amendment to the Clean Air Act is the only part of the act designed specifically to help alleviate acid rain. However, this provision was developed mainly to meet the reciprocity requirement of Section 115 of the U.S. Clean Air Act; actual implementation of this provision, bypassing the authority of the provinces, is unlikely.

The new automobile emission standards will result in the same level of compliance for Canadian vehicles as that currently used in the United States. This is expected to reduce overall NO_x loading by 45 percent by the year 2000.

Although the Clean Air Act provides only a limited legal basis for the federal government to reduce acid rain-causing emissions, no changes in the act are currently planned. Rather, as explained in the next section, the federal government is working cooperatively with the provincial governments to reduce emissions from stationary sources.

Federal Acid Rain Policies

Because, for legal or political reasons, the existing air pollution laws limit federal control of SO_2 and NO_x emissions in Canada, the federal government uses three strategies to help solve the acid rain problem: research and monitoring, encouragement of provincial action, and encouragement of U.S. action.

Research and Monitoring

Between 1980 and 1984 the federal government committed $41 million (Cn) to fund research on acid rain.[32] On a per capita basis this represents several times more than that spent by the United States during the same period. Current expenditures equal $30 million annually, with more than $18 million coming from the federal government.[33] The research is conducted by several different departments, including the Department of the Environment, Health and Welfare Canada, and the Department of Energy, Mines, and Resources. Research projects include studies of the effects of sulfuric acid when added to healthy lakes, studies of the impacts of acid rain on crop development and forest growth, and epidemiological studies of the effects of sulfate particles on people with respiratory illnesses, as well as projects to develop better atmospheric modeling techniques.[34] Moreover, Environment Canada has established several atmospheric monitoring systems to provide data on the chemical composition of precipitation. The major system—the Canadian Network for Sampling Precipitation—consists of fifty-four stations; other networks such as the Air and Precipitation Monitoring Network are smaller. The federal government believes that the information gained from this research may provide the basis for new control strategies.

Encouraging Provincial Action

The federal government encourages provincial governments to apply more rigorous standards against Canadian emitters of acid rain precursors.[35] This has been accomplished both through private meetings between the federal minister of the environment and his provincial counterparts and through the development of federal-provincial task forces.[36] For example, the Federal-Provincial Steering Committee on Control Strategies for Long Range Transport of Atmospheric Pollution was established to "identify, develop, and evaluate alternative abatement strategies."[37] The Ontario-Canada Task Force had the mandate to investigate control options and socioeconomic impacts for reducing emissions by INCO, Ltd., and Falconbridge Nickel Mines.[38]

Federal and provincial governments agreed in 1982 that the Canadian environment would be protected from the effects of acid rain if acid deposition were reduced to twenty kilograms per hectare per year (eighteen pounds per acre per year). In March 1984, when hope for a speedy bilateral action on the Canadian proposal for a joint 50 percent reduction in U.S.-Canadian SO_2 emissions had vanished, Canadian federal and provincial officials announced independent action to cut Canadian industrial emissions by 50 percent in the next decade. By February 1985 federal and provincial environmental ministers had agreed on specific emissions ceilings for each of the seven provinces east of the Saskatchewan-Manitoba border (table 6-1). The program contains specific emissions reduction targets while allowing industrial emitters to choose the means to achieve them. It is estimated that the fully implemented program will cost $500 million (Cn) annually.[39] Ontario and Quebec, which together account for 75 percent of the reduction, have already enacted regulations to implement the program.

Although the success of this effort depends on provincial regulation, the federal government has provided some financial assistance to facilitate implementation. It has allocated $25 million for the development of new pollution control technologies for nonferrous smelters; $150 million for their installation and operation; and an additional $30 million for an ongoing research program to monitor the efficacy of the abatement effort.[40]

Table 6-1.   Sulfur Dioxide Reduction Targets by Province
(thousands of metric tonnes)

Province	1980 Base case	1994 Commitments	Percentage reduction
Manitoba	738	550	25
Ontario	2,194	885	60
Quebec	1,085	600	45
New Brunswick	215	185	26
Prince Edward Island	6	5	17
Nova Scotia	219	204	7
Newfoundland	59	45	24
Total	4,516	2,474	45

Source: Environment Canada, "Stopping Acid Rain," pamphlet.

Encouraging U.S. Action

The most significant element of acid rain policy by the federal government has been its efforts to encourage the United States to reduce its SO_2 and NO_x emissions. The Canadian government strongly believes that 50 percent of the acid rain deposition in Canada is caused by emissions in the United States. Efforts to solve the acid rain problem, therefore, should come from *both* countries. At a conference in Buffalo, New York, Canadian Secretary of State for External Affairs Mark MacGuigan summarized the federal government's view: "In light of this legislative action [the amendment to the Clean Air Act], and the actions taken to begin controlling Canadian pollution sources, Canadians now expect the United States to demonstrate the same degree of concern to address the problem. In short, we in Canada are convinced that we cannot resolve acid rain ourselves. We urgently need the cooperation of the United States."[41]

The federal government has encouraged U.S. action in two ways. The first has been to try to increase awareness of the acid rain problem among U.S. legislators and the public in order to encourage legislative action. Canadian embassy officials have testified at hearings before U.S. congressional subcommittees investigating the acid rain issue to emphasize both the urgency of the problem, and the need for the United States to reduce emissions of sulfur dioxide and nitrogen oxides in order to solve the problem.[42] In the early 1980s the Canadian government developed an intensive information distribution campaign in the United States in order to increase awareness of the acid rain problem. The campaign included the sponsoring of conferences in the United States on acid rain; extensive speaking engagements by upper-level Canadian officials to U.S. audiences; and the distribution of leaflets, TV film clips, fact sheets, and slide shows to emphasize that acid rain is a serious problem. The campaign tried to encourage the public to pressure their representatives to take action.[43] In discussing this policy, journalist Lois Ember noted that: "Canadians understand that increased public awareness and sensitivity to the issue will most likely translate to increased pressure on members of Congress to take some action."[44] The more controversial aspects of the public awareness campaign have been stopped since the election of a more conservative government in 1984.

Second, Canada seeks a bilateral air quality agreement with the United States aimed at reducing emissions of acid rain precursors, which it perceives as the only real solution to the acid rain problem.[45] On August 5, 1980, Canada and the United States signed a Memorandum of Intent on Transboundary Air Pollution. Chapter 3 examined the Memorandum of Intent and the difficulties encountered in negotiations between the two countries since it was signed. In 1982 the Canadian federal government submitted to the United States a proposal to reduce Canadian SO_2 emissions by 50 percent by 1990, contingent on parallel action by the United States. This proposal was based on an agreement reached with the eastern provinces that acid deposition should be reduced to twenty kilograms per hectare per year in order to protect moderately sensitive ecosystems.[46] The Canadian government estimates that, although the costs of achieving a 50 percent reduction may, overall, be higher for the United States, the costs per capita would be three to four times greater in Canada.[47] The U.S. negotiators, however, rejected this proposal.[48]

In another strategy Canada hosted an international meeting on acid rain in March 1984. The environment ministers of Canada and nine western European countries signed a declaration to reduce sulfur emissions at least 30 percent in the coming decade. It was hoped that this ploy would put pressure on neighboring countries such as the United States and Great Britain, who declined to endorse the proposal. However, further discussions are planned, and another eight countries have agreed to join the "30 Percent Club."[49]

Summary and Conclusions

Because the federal government is unable to regulate emissions of acid rain precursors in Canada directly, and because it cannot control emissions of SO_2 and NO_x from the United States, it has concentrated its efforts on encouraging both the provincial governments and the United States to take actions to solve the acid rain problem. To do so it funds research on acid rain and engages in other activities in order to provide the provinces, and the United States, with information on the effects of acid rain and on control strategies. In this view, the solution to the acid rain problem lies with the provinces and the United States.

Provincial Governments

This section examines the handling of the acid rain problem by the governments of two Canadian provinces, Ontario and Quebec. These two provinces are examined in detail because: (1) they contain large SO_2-emitting sources and therefore contribute significantly to the acid rain problem; and (2) they suffer from the effects of acid rain and thus have an incentive to solve the problem.

The Legal Basis for Controlling Acid Rain in Quebec and Ontario

The Ontario Environmental Protection Act of 1971 is the major legislative mechanism for controlling air pollution in Ontario, and hence for controlling emissions of acid rain pollutants. The act empowers the minister of the environment to conduct meteorological measurements and studies on the quality of the natural environment, to establish monitoring programs, to convene conferences on environmental problems, and to gather, publish, and disseminate information relating to pollution.[50] The act also establishes the regulatory mechanisms for controlling air pollution in Ontario. It empowers the department to issue certificates of approval and control orders, and it gives authority to the lieutenant governor in council (the provincial governor acting with the consent of his cabinet) to establish regulations to control the emission of air contaminants.

The provisions of Quebec's Environmental Quality Act are similar to those in Ontario's Environmental Protection Act. Under the Environmental Quality Act, the minister of the environment is empowered to conduct research and studies on the quality of the environment, to establish monitoring programs throughout Quebec, and to publish statistical data on the quality of the environment.[51] Moreover, the Environmental Quality Act empowers the directeur de Services de protection de l'environnement (hereinafter referred to as the director), who is appointed by the minister, to issue certificates of authorization and cessation orders, and it empowers the government to establish regulations to control the emission of air contaminants. Because the provisions of both acts are so similar, they will be discussed together.

Certificate of Approval/Authorization   Both acts provide that no one may construct or alter a plant from which a contaminant may be discharged into the environment, unless he has obtained a certificate of approval (in Quebec, a certificate of authorization) from the director.[52] The director has the authority to require any alteration in the plant —including installation of pollution control technology—to be resubmitted for approval in order to ensure that the emissions from the plant will not pose a hazard to the health or safety of human beings or impair the quality of the natural environment.[53] If the polluter disagrees with the requirements set forth by the director, he has the right to appeal the decision. In Ontario the Environmental Appeal Board has the power to alter or revoke the conditions for approval. This decision may be further appealed to the courts on questions of law or to the minister of the environment on questions other than law.[54] In Quebec the decision may be appealed to the Commission Municipale du Quebec, which has similar powers.[55]

After acid rain has been shown to cause damage to the natural environment, the Ontario Ministry of the Environment and the Quebec Department of the Environment could, on a case-by-case basis, require new SO_2 and NO_x polluters to install pollution control equipment as a condition for plant construction. However, use of this provision to control acid rain-causing emissions is difficult. If the polluter takes advantage of the right of appeal, issuing a certificate to control SO_2 and NO_x emissions may prove to be costly and time consuming. To avoid these problems the Ontario Ministry of the Environment has, in practice, sought to negotiate the conditions for approval with individual plants.[56] Moreover, the usefulness of this provision for reducing acid rain-causing emissions is further constrained by the definition of the *natural environment* in the Ontario Environmental Protection Act as the "air, land, and water or any combination or part thereof, of the *Province of Ontario*" (emphasis added).[57] Strictly speaking, damage to the environment outside Ontario, such as that resulting from the long-range transport of SO_2 and NO_x, cannot be used by the director to insist on the installation of pollution control equipment before issuing the certificate of approval. In practice, however, damage to the environment outside Ontario has been implicitly included in designing the conditions for approval. But the requirement for major SO_2 and NO_x emitters to install pollution control equipment before receiving a certificate of approval applies only to new or modified plants in Ontario. However, existing plants may not need to make major alterations for many years, and do not need to install pollution control equipment until that time. As a result, there may be a long delay between the implementation of the pollution control requirement and an actual reduction in sulfur and nitrogen emissions. In contrast, Quebec's certificate of authorization applies to all sources, not just new or altered plants.

Control/Cessation Order Under the two provincial acts, the director may issue a control order (a cessation order in Quebec) upon the finding that a contaminant has been emitted into the environment at a level that exceeds the maximum permissible amount stated in the regulations, or at a level that may cause harm to human health or impair the quality of the natural environment.[58] The order may be a requirement to reduce or stop emissions or, more specifically, it may require the polluter to install specific pollution control equipment. The control (cessation) order is source specific and thus has the advantage that it can be tailored to each specific problem. The order, similar to the certificate of approval (authorization), can be appealed by the polluter to the Environmental Appeal Board in Ontario (the Commission Municipale du Quebec). The provinces can issue control orders to major SO_2 and NO_x emitters to require reduction of these emissions, or specifically to require installation of pollution control equipment.

However, like the certificate of approval, effective implementation of this policy is difficult. First, control orders can be appealed by the polluter. Consequently, the issuing of control orders, if appealed, may prove to be costly and time-consuming. Authors David Estrin and John Swaigen note this problem with a control order in Ontario, pointing out that "the Ministry has the power to impose stringent conditions in a control order, but the polluter has the right to appeal to the Environmental Appeal Board and to the courts.... such appeals can delay the effect of the control order for as much as two years, and it is still possible that the Environmental Appeal Board or the courts will refuse to grant the control order or water it down."[59]

This problem in using a control order to reduce SO_2 emissions in

Ontario was recognized by the Canada-Ontario Task Force, which, in evaluating legislative mechanisms to reduce SO_2 emissions in the Sudbury basin, concluded: "Due to the complexity of the technical options, the extended time period in which implementation would take place and the inherent potential for delay. . . . this mechanism is not appropriate for the task at hand."[60]

Regulation   Both the Ontario Environmental Protection Act and the Quebec Environmental Quality Act provide that the lieutenant governor in council (of each province) may make regulations controlling emissions into the natural environment from any source of contamination.[61] The acts do not require the regulation to be based on any particular criteria such as health effects and damage to the environment. However, the regulations are subject to judicial review to determine if they exceed the intent of the act. In contrast to the control order, a regulation cannot be appealed to the Environmental Appeal Board in Ontario or to the Commission Municipale in Quebec.[62] Consequently, issuing a regulation can provide an effective means of acid rain control through the control of emissions of SO_2 and NO_x from specific sources.

In addition to source-specific regulations, the acts also authorize regulations establishing maximum permissible ambient concentrations of any class of contaminants throughout the province.[63] Quebec has developed maximum permissible concentrations based on the federal National Ambient Air Quality Objectives. Quebec's maximum permissible concentration for SO_2 emissions is 52 micrograms per cubic meter per year, which is in the "acceptable" range of the federal guidelines. Because these regulations take the form of maximum permissible ambient concentrations, they do little to reduce the overall loading of acid rain-causing pollutants into the environment.

In Ontario the regulations technically do not take the form of maximum permissible ambient concentrations, but rather are calculated as one-half-hour concentrations at the "point of impingement" (the point of the closest receptor). The point of impingement regulations are designed to satisfy local ambient air quality criteria.[64] In the case of SO_2, the point of impingement limit is 830 micrograms per cubic meter in a half-hour interval, but this is designed to achieve an ambient concentration limit of 55 micrograms per cubic meter as a maximum annual level. This ambient level falls within the "acceptable" range of the federal National Ambient Air Quality Guidelines. Thus the maximum permissible amount is essentially an ambient standard, and again does little to reduce the overall loading of acid rain-causing pollutants into the atmosphere.

Summary   The provinces of Quebec and Ontario have several tools available to control the emission of SO_2 and NO_x into the environment: the certificate of approval/authorization, the control/cessation order, and the regulation. The effectiveness of the certificate of approval/authorization and the control/cessation order in reducing emissions is limited by delays in implementation and appeals. In Ontario restrictions can only be implemented if damage to the natural environment within the boundaries of Ontario is likely to occur or has occurred; damage to the natural environment outside Ontario is not sufficient. Thus, the most effective regulatory tool for controlling acid rain-causing emissions appears to be the issuance of source-specific regulations by the lieutenant governor in council. A regulation is not subject to statutory appeal and therefore it is a powerful tool. At the same time, regulations establishing a maximum permissible level of a contaminant are based on ambient criteria that do not address overall loading of acid rain-causing emissions.

Acid Rain Policies of Ontario and Quebec

In response to the acid rain problem, both Quebec and Ontario have adopted three major policies within their legal powers to help solve the problem: research and monitoring; emissions reduction of major SO_2 sources; and encouragement of U.S. action. Although, in general, their major policies are the same, their specific methods of implementing these policies have differed.

Research and Monitoring   Established in 1979, the Acidic Precipitation in Ontario Study (APIOS) is a coordinated effort to investigate the causes and effects of long-range transport of air pollutants. The program includes both scientific and socioeconomic studies, directed by an interbranch or interministerial working group. More than six thousand lakes have been sampled in order to evaluate the extent of acidification. The greatest concentration of acidified lakes was found in the Sudbury area. However, significant improvements can be correlated with reductions in SO_2 emissions. The effects of emissions reductions in both Canada and the United States are being investigated with the aid of a long-range transport model installed in the University of Toronto's CRAY XMP computer.[65]

Quebec has set up a network of forty-six monitoring stations, mainly in southern Quebec, to monitor wet and dry deposition.[66] Current research projects include an in-depth inventory of lakes and their environment, studies on the effects of acid rain on aquatic life, socioeconomic impact studies, and research on alternative control strategies, such as a pilot project to convert SO_2 emissions to magnesium sulfate.[67]

Emission Reduction of Major SO_2 Sources   Beginning in 1980 the province of Ontario sought to reduce acid deposition by issuing regulations to two of its major sources of SO_2, INCO, Ltd., and Ontario Hydro Electric Company. These two sources combined accounted for 70 percent of Ontario's 1980 emissions of SO_2.[68] These regulations were followed in 1985 by more stringent measures that became part of a nationwide attempt to reduce SO_2 emissions. Although these regulatory actions are aimed at reducing the acid rain problem, they are also designed to demonstrate to the United States Ontario's seriousness in wanting the acid rain problem solved.[69]

INCO is the world's largest nickel producer and North America's largest single source of SO_2 pollution. In September 1980 the lieutenant governor in council issued a regulation to INCO's plant in Sudbury requiring it immediately to reduce its emissions from 3,600 tons of SO_2 per day[70] to 2,500 tons per day, and to further reduce its emissions to 1,950 tons per day by 1983.[71] Because Ontario issued a regulation instead of a control order, INCO is legally bound to comply with the order and cannot appeal the decision. Ontario Minister of the Environment Keith Norton, referring to the decision to issue a regulation to INCO, stated that "the government chose to implement these levels, not by the usual control order procedures which could have been stayed by company appeals, but rather, by using a government regulation which does not carry with it the regular appeals mechanism."[72]

INCO achieved the immediate reduction to 2,500 tons per day by cutting back on its production.[73] In 1982 INCO was shut down entirely because of the depressed nickel market and a strike. Changes in the production process were made during this time, and when the smelter resumed production in late 1983 its daily SO_2 emissions averaged about 1,900 tons.

The Ontario government also issued a regulation to Ontario Hydro, the province's utility company, requiring it to reduce substantially the SO_2 and NO_x emissions from its coal-fired electric generating plants. Ontario Hydro is the second largest source of SO_2 within the province, and it is the largest industrial source of NO_x. Its coal-fired plants

produce about 20 percent of Ontario's total emissions of these pollutants. In January 1981 the province issued a regulation to Ontario
Hydro requiring it to reduce SO_2 and NO_x emissions from the 1982
level of roughly 607,000 tons per year to 450,000 tons in 1985, and
down further to 300,000 tons in 1990, of which no more than 260,000
tons could be SO_2. This would represent a 43 percent reduction in
aggregate sulfur dioxide and nitrogen dioxide emissions from Ontario
Hydro by 1990. The regulation was, again, nonappealable, and Ontario
Hydro was required to meet these regulations regardless of any
increase in power demand.[74] As part of its response, Ontario Hydro is
in the process of shifting much of its power production to nuclear
plants—a viable strategy in Canada due to the absence of strong
opposition to nuclear power. Coal-fired plants will increasingly be
restricted for use during peak demand periods.

Following these initial attempts to reduce emissions by targeting its
two main sources, Ontario devised a more comprehensive program
in response to nationwide efforts to control acid deposition. In
December 1985 Ontario announced its Countdown Acid Rain program. As a result of negotiations between the federal and provincial
governments to reduce acid deposition by 50 percent, Ontario had
agreed to reduce its emissions of SO_2 from the 1980 base case of 2,194
kilotons to 885 kilotons by 1994. With 80 percent of the emissions
originating from four corporate sources, nonappealable regulations
for staged reductions were developed for each of these sources. The
regulations require the companies to file biannual reports outlining
abatement methods and technologies chosen, and detailing the socioeconomic implications.[75]

Reduction targets set for 1986 in the early 1980s for INCO and
Ontario Hydro were revised upward and additional reductions by 1994
were mandated. For INCO the original 1986 target of 728,000 tons per
year has been changed to 685,000 tons per year, going to 265,000 tons
per year in 1994. In addition, INCO is required to develop cost estimates for a more stringent objective of 175,000 tons per year. Ontario
Hydro must reduce its 1986 target from 390,000 to 370,000 tons per
year, with a 1994 target of 175,000 tons per year. Both the Sudbury
smelter of Falconbridge and the Wawa operation of the Algoma Steel
Corporation are required to reduce emissions by one-third between
1986 and 1994. Other small reductions will also be realized from
specific control strategies on smaller point sources, a new general
boiler regulation, and a shift in the relative price of fuels.[76]

In the early 1980s Quebec began to control SO_2 emissions with the
goal of reducing emissions by 27 percent by 1985.[77] This was to be

accomplished in two ways. The first required reductions of SO_2 emissions from its major sources, notably the Noranda and Gaspe mines. Noranda Mines is Quebec's largest source of SO_2 emissions, accounting for more than 50 percent of total anthropogenic emissions of SO_2 in Quebec, and it is the second-largest source of SO_2 in all of Canada.[78] In 1983 Noranda Mines, under orders from the Department of the Environment, submitted to the department an assessment of the economics of proceeding with the development of a process to produce magnesium sulfate from SO_2 and asbestos tailings. At that time the Department of the Environment was considering a proposal to issue a control order to Noranda Mines to reduce its SO_2 emissions by 40 percent by 1985,[79] and to Gaspe Mines to increase its recovery of sulfur from 50 percent to 65 percent at its smelter and acid plant.[80] By 1984 a 31 percent reduction had been achieved.[81]

In February 1985 Quebec modified its Environmental Quality Act and became the first province to pass an acid rain control program that had the force of law. As a result, sulfur emissions will be cut by 45 percent overall by:

   reducing emissions from existing continuous reactor smelters by 50 percent by 1990 as compared with 1980 levels;

   limiting emissions at all other smelters to 275 kilograms of SO_2 per ton of mineral concentrate;

   requiring all new smelters to control 95 percent of sulfur contained in the concentrate or ore.[82]

Encouragement of U.S. Action   Both Ontario and Quebec have attempted to encourage the United States to reduce its SO_2 and NO_x emissions. The two provinces agree with the federal government that much of the acid rain deposition in Canada is caused by emissions in the United States. Ontario and Quebec have encouraged U.S. action in a variety of ways.

Both provinces have taken legal action to encourage the U.S. EPA to prevent increases in the current level of SO_2 emissions. In March 1981 the EPA was petitioned by several midwestern states to permit increases in the emissions limits of twenty coal-burning power plants. Most of these states are located in the Ohio valley, and they are suspected to be major contributors to the acidic deposition problem in Ontario and Quebec. In response to these petitions, the Ontario Ministry of the Environment intervened in support of petitions filed by New York and Pennsylvania that opposed the proposed relaxation in standards. Ontario urged the EPA to break with its tradition of considering only local effects of such a revision, and proposed that EPA evaluate the cumulative effect that the increase in SO_2 emissions might have on Ontario.[83] Similarly, Quebec submitted a brief to the EPA in support of the petitions filed by New York and Pennsylvania. The brief stated Quebec's concern about the acid rain problem and urged the EPA to give all due consideration to Quebec's objections to increasing the SO_2 emissions levels.[84] In 1981 Ontario filed three other interventions in response to similar proposals. These legal actions served not only to influence U.S. actions that might have affected Ontario and Quebec, but also increased awareness among the U.S. public of the acid rain problem. The Ontario Ministry of the Environment wrote in regard to their legal interventions that "through these activities we have generated important support among the U.S. media and public."[85] The EPA did not act in response to the petitions by the northeastern states and the Canadian provinces. In March 1984 six northeastern states filed suit against EPA for failure to curb acid rain. However, the EPA administrator, in a letter to the attorney general of Maine, expressed doubt that the U.S. Clean Air Act could be used for this purpose. The act, in his opinion, was only concerned with "ambient standards, not with acid rain."[86] Subsequently, the courts upheld the position of EPA, on appeal (see chapter 9).

Ontario and Quebec have also encouraged U.S. action in an indirect manner. Both provinces have worked with the Canadian federal government and the other eastern provinces to develop a bilateral transboundary air quality agreement between the United States and Canada. Ontario and Quebec and the other eastern provinces took an active role in developing the proposal to the United States that would reduce eastern Canada's SO_2 emissions by 50 percent by 1990, contingent upon parallel action by the United States.[87] As noted before, this proposal was rejected by the U.S. negotiators. Because the provincial governments are responsible for implementing such provisions, the federal government would not have been able to make this proposal without their consent.

Moreover, Ontario has sought to encourage U.S. action by adding to the federal government's efforts to increase the awareness of the acid rain problem among U.S. legislators and the U.S. public. Representatives from the Ontario Ministry of the Environment have testified before U.S. congressional committees investigating the acid rain problem, in order to emphasize the urgency of the problem and the need for the United States to reduce its emissions of sulfur dioxide and nitrogen oxides.[88] In addition, the Ontario Ministry of the Environment has, in cooperation with the federal government, sponsored on-site briefings for congressional aides, senatorial representatives, and the media to emphasize the importance of the problem. The ministry, in discussing these actions, states that: "We are convinced that all of these communication activities have done much to awaken the American public, including the news media and the U.S. legislators, to the acute problem of acid rain."[89]

Finally, Quebec has joined with individual states in the United States, especially in the Northeast, in an effort to encourage reduction of U.S. emissions. Thus far, Quebec has concluded agreements with New York and Vermont. The agreements recognize the importance of joint research efforts and the exchange of information on the acid rain phenomenon. Moreover, the parties agreed to form a common strategy to influence the U.S. Congress to implement a reduction in SO_2 emissions.[90] In this way Quebec has sought to ensure that "Quebec's firm stand against an increase in the transboundary air pollution transport be strongly felt and loudly heard in Washington."[91]

Summary

The Canadian federal government does not have the legal or political authority to regulate the emission of acid rain precursors; therefore, it has encouraged the provincial governments and the U.S. government to take action to control these emissions. Seven eastern provinces have agreed to apportion reductions of SO_2 in order to achieve an overall reduction of 50 percent. Of this, 75 percent will be realized by Ontario and Quebec. Ontario has issued nonappealable regulations to four of its major sources of SO_2, requiring them to reduce emissions. Quebec has developed a legally enforceable acid rain control program. Both provinces believe that the solution to the acid rain problem lies not only with reductions in the SO_2 emissions within Canada, but also with reductions of SO_2 emissions in the United States. Consequently, both provinces have adopted policies of encouraging the U.S. government to reduce its emissions. Finally, both provinces have developed monitoring networks and have funded acid rain research in order to provide information for themselves and the United States on the effects of acid rain and on control strategies.

Conclusions

Is Canada Doing Enough at Home?

In the past Canada has been criticized for pressuring the United States to take action to reduce sulfur and nitrogen emissions before "getting its own house in order." Critics have charged that Canadian air pollution laws are weak and that the Canadian government does not have the authority to control SO_2 and NO_x emissions within its boundaries. The analysis presented here shows that, despite the fact that the federal government is constrained legally and politically from directly regulating sulfur and nitrogen emissions in Canada, it has succeeded in coming to an agreement with the provincial governments to achieve a 50 percent reduction in SO_2 emissions from stationary sources. Ontario and Quebec, which together account for 75 percent of eastern Canada's emissions, have already implemented programs to attain these reductions. In addition, the federal government has taken steps to bring the standards for automobile emissions into compliance with U.S. standards.

Is Canada Doing Too Much or Too Little to Influence U.S. Policy?

Canada has also been criticized for trying to influence U.S. policy. Critics have charged that Canada's active involvement in the U.S. political process is creating resentment among U.S. legislators and the U.S. public, and is detrimental to solving the problem of acid rain. Several questions are raised: Should Canada encourage U.S. action? If so, at what point will they have "gone too far"? Has that point been reached yet? Because much of Canada's acid rain results from sulfur emissions from the United States, and because bilateral air pollution negotiations have made little progress, Canada's need to encourage U.S. action is real. It is clear, however, that at some point Canadian participation in the U.S. political process will be viewed as interference by the U.S. population; but exactly when this point will occur is difficult to judge. A wise strategy for Canada would be to emphasize educating the U.S. public to the problem of acid rain, but not to become too active a participant in the U.S. political process.

Chapter Seven
The U.S. Policy Response to Acid Rain

ıııııı

Introduction

The U.S. policy response to acid rain, like the U.S. policy response to previous environmental problems, has been reactive rather than active. Before the 1980 presidential elections the federal government's involvement in acid rain policy was mostly in response to Canadian concerns and focused on negotiations with the neighbor to the north but had not begun to approach a coherent policy for acid rain control.[1] Under pressure from the Canadian government and environmental groups, several federal agencies were conducting research, and an Interagency Task Force on Acid Rain had been established to provide a better scientific basis for possible control measures. However, serious efforts to control acid rain did not begin until it became necessary to reauthorize the Clean Air Act. The reauthorization process, which began in 1981, has been long and difficult because of competing expectations for the outcome among policymakers and interest groups.

As of spring 1988, efforts to reauthorize the act, which expires in August 1988, are still in progress. There is considerable pressure for legislative action. The interest groups discussed in chapter 5 continue to play an active role in the process, and it remains to be seen whether the Clean Air Act will become an effective vehicle for acid rain controls. This chapter examines the U.S. policy response to acid rain, including the activities of executive agencies in relation to acid rain. The discussion also focuses on the major policy initiatives in both houses of Congress, as well as initiatives at the state level.

This chapter was written by Barbara Britton, Tom Albin, and Steve Paulson.

The Role of the Clean Air Act in Acid Rain Policy

Because acid rain poses an ecosystem-level threat and crosses political boundaries, it must be addressed at the highest level of government and through the most comprehensive available policy tools.[2] In the United States this means the Clean Air Act. No other environmental legislation has addressed the regulation of air pollutants in such a broad framework. But the Clean Air Act as it stands is inadequate to solve the policy problems posed by acid rain. For the Clean Air Act to work, Congress must adapt the act to the peculiarities of acid rain.

Background on the Act

In a 1970 environmental message to Congress, President Richard Nixon stressed the need for immediate pollution control legislation. Congress shared that view and, under the leadership of the Senate Committee on Environment and Public Works, enacted the Clean Air Act—the first of a series of broad federal environmental protection programs of the 1970s. In enacting this legislation, Congress recognized that all Americans are entitled to live in a healthy and clean environment and that an effective, coordinated government effort was necessary to achieve that goal.

Since the passage of the act, the century-long trend toward deterioration of the nation's air has abated, and air quality has actually improved. The National Commission on Air Quality (NCAQ), a bipartisan committee established by Congress to review the Clean Air Act, found significant improvements in air quality over the ten years since the act was passed. Between 1974 and 1978, for example, the commission found that the number of days which could be classified as unhealthy in the nation's twenty-three major metropolitan areas declined by 18 percent. Between 1973 and 1978 the annual average concentration of SO_2 decreased by 20 percent. These improvements in air quality have produced substantial benefits for the country. The NCAQ cited studies showing that Americans have suffered fewer illnesses and fewer premature deaths, and have enjoyed lowered health-care costs because of the Clean Air Act. Beyond these health benefits, Americans have enjoyed important secondary benefits, including reduction in damage to crops and lower maintenance costs for buildings.

These improvements have not been without costs, nor were they expected to be. Overcoming decades of environmental neglect and abuse has required a major commitment of resources. However, the aggregate negative economic impact of the act (including higher prices, increased unemployment, and decreased productivity), in the words of the NCAQ, "has not been significant and is not expected to be significant through ... the mid 1980s." Moreover, when economic impacts that are not reflected in conventional market transactions —such as effects on health and the environment—are considered, the overall economic effects of the Clean Air Act are clearly positive.

However, the Clean Air Act as it stands does not solve the nation's air quality problems. Many Americans still live in areas where air pollution presents significant health hazards. In early 1981, 103 of the country's 105 urban areas with populations over 200,000 exceeded one or more of the ambient air quality standards. In 1980 more than 140 million Americans lived in areas where the ozone standard—a standard set to protect public health—was not being met. Similarly, over 73 million citizens lived in regions where the particulates standard was exceeded, and 62 million faced unhealthy concentrations of carbon monoxide.

The Senate Committee on Public Works and the House Subcommittee on Health and Environment have been reviewing the act, which expired on September 30, 1981. The issues involved are basic yet complex: How clean must the nation's air be in order to protect public health and welfare? How costly and restrictive has the act been? Are the provisions the most efficient measures available? The Ninety-seventh Congress failed to reach a conclusion on the reauthorization, as did the Ninety-eighth and Ninety-ninth Congresses. A part of the reason for this failure is the debate over the appropriate way to address the acid rain issue.

The Clean Air Act and Acid Rain

Because the Clean Air Act was framed before acid rain was recognized as a danger, it does not adequately address the acid rain problem. Moreover, in several important ways the act has contributed to the problem. The State Implementation Plan (SIP) process and the National Ambient Air Quality Standards (NAAQS), the key mechanisms for limiting emissions, focus on ground-level concentrations of targeted pollutants. These mechanisms are not capable of addressing regional spillovers of SO_2 and NO_x stimulated by the increased use of tall stacks. By concentrating on ambient air quality, the act encouraged the use of tall stacks to dispose of pollution so that local air quality standards could be met. Since 1970 electric utilities have built almost three hundred stacks, and the average height has increased from two hundred feet in 1960 to over six hundred feet today. At these heights winds are stronger and more constant than they are closer to the ground. Consequently, more pollutants are staying aloft longer and traveling farther than before. The longer SO_2 and NO_x stay aloft, the greater is the likelihood that they will be converted into acids. Amendments to the Clean Air Act in 1977, recognizing dangers associated with long-range transport of pollutants, imposed height limitations on industrial stacks.

In addition, the act does not require previously existing sources of emissions to meet the New Source Performance Standards (NSPS), which require all coal-fired utilities to remove 70–90 percent of the fuel's sulfur. These standards reduce the emissions levels of new plants to about one-seventh of previously existing plants' levels. Moreover, allowing each state to regulate itself under the SIPs has resulted in uneven standards. For example, Ohio allows thirty times the amount of SO_2 emissions permitted in Connecticut. Consequently, the act permits clean-plant states in New England and elsewhere to be dirtied by states with lesser air standards.

The act does contain several provisions affecting the control of acid rain, however. In order to see how these provisions affect acid rain, we will discuss each briefly.

National Ambient Air Quality Standards   The NAAQS are standards for local air quality that depend on measures of pollutant concentrations at ground level. Thus, they do not directly control the amounts of pollutants emitted, and they encourage the dispersion of pollutants as discussed above. The act provides for primary standards to protect "public health" and for secondary standards to protect "public welfare." These standards, unlike NSPS, apply to sources in existence before the law was enacted in 1970. Secondary standards were supposed to have been more stringent than primary standards, but no attainment deadline was set for these standards, effectively rendering them useless. NAAQS might have been particularly relevant to acid rain control because Congress intended them to protect against damage to buildings, materials, structures, animal life, and terrestrial and aquatic ecosystems. But they are either lacking entirely or are no more stringent than primary standards, and they are not enforced. Currently, these standards apply only to sulfur and nitrogen oxides. EPA has not established sulfate and nitrate standards, claiming that there are no conclusive research findings upon which to base them. However, it has been argued that Section 108 of the act requires that EPA list and promulgate standards not only for sulfur dioxide, but also for sulfur compounds derived from them. This issue formed the basis of a lawsuit brought against EPA by seven states and four environmental organizations.[3]

Control of Pollution from New Sources   The Clean Air Act controls new mobile and stationary sources more stringently than older ones, in order to lessen the costs of control for older sources while improving air quality over time as new plants are built. Several sections of the Clean Air Act apply specifically to new sources. This reasoning has not worked out quite as planned, however.

The federal government began controlling automobile emissions in 1965[4] and required automobile manufacturers to install pollution control devices under the Clean Air Act of 1970. Automobile manufacturers have since been required to install the most stringent state-of-the-art control technology. Although automobiles are one of the primary sources of nitrogen oxides, they do not represent a likely target for controlling acid rain because the currently used technology cannot yield significantly higher levels of control.

New stationary sources of sulfur and nitrogen pollution are controlled under the NSPS, the provisions for Prevention of Significant Deterioration (PSD), and Nonattainment Provisions. The most important requirement is for installation of control technology during construction of new sources. The NSPS require new sources to install the best technologically practicable and economically feasible controls for nitrogen and sulfur oxides. Such control technology, however, is expensive, and the NSPS provide an incentive for plant owners to keep older, polluting facilities operating longer than ordinarily expected. Significant decreases in emissions of nitrogen and sulfur oxides (assuming a constant level of economic activity) are not expected until 1990, when existing facilities reach their maximum life span and new facilities must be built to replace them.[5] PSD requirements are designed to place stringent controls on sulfur oxide- and particulate-emitting new sources located in regions where air quality is better than that required by law. Although these provisions prevent industry from moving sources of pollution from dirty to clean areas, they do not result in actual emissions reductions. Although PSD requirements can be useful in preventing increases in acid rain, they will not help in its abatement.

Nonattainment Provisions are designed to reduce pollution levels in areas where air quality does not meet requirements of current law. The level of control required by these provisions could lead to reductions in nitrogen and sulfur oxide pollution and help to alleviate the acid rain problem. However, these provisions are applicable only to sources located in nonattainment areas, and relatively few areas have not met the standards.[6]

Prevention of Interstate Pollution   The Clean Air Act addresses interstate pollution through provisions in the State Implementation Plans, in the permit requirements for PSD, and in the Nonattainment Provisions. SIPs detail strategies for implementation of the Clean Air Act by state and local governments and are subject to EPA approval. The plans address such items as individual source emission limits, control technologies to be used in meeting these limits, methods of monitoring and enforcement, and assurance that the proposed measures will prevent interstate pollution. In defining SIP requirements, the Clean Air Act lists specific requirements for controlling interstate pollution, including the obligation for states to ensure that pollution generated within their borders does not interfere with another state's ability to meet national ambient air quality standards or other state air pollution requirements. The act provides legal authority to state or local governments to petition EPA to take action when SIP provisions against interstate pollution have been violated.

There are three problems associated with using SIP provisions and Clean Air Act provisions to control interstate transport of acid rain-causing air pollution. First, these provisions can only be used after standards have been violated. Second, after release in the atmosphere, sulfur and nitrogen oxides are transformed into sulfates and nitrates; by the time they reach another state they are not in a form that is controlled by existing ambient air quality standards. The third problem arises with the burden-of-proof requirements. Current Clean Air Act provisions place the entire burden of proof of violations on EPA, which must identify a specific source of pollution before it can take action. This is impossible with current detection methods because pollution can be transported hundreds of miles from its source.

PSD provisions and Nonattainment Provisions contain requirements for new sources to show that they will not cause interstate pollution, and they place the burden of proof on sources rather than on the EPA. But, again, these programs can help to prevent increases in acid rain precursor emissions, but not to reduce present levels of pollution.

International Air Pollution   Section 115 of the Clean Air Act provides a mechanism for controlling international air pollution by allowing a foreign country to petition EPA to force a state whose pollution causes adverse health and welfare effects in the foreign country to amend its SIP to stop the transboundary pollution. Before foreign entities are allowed to use the U.S. courts to enforce Section 115 of the Clean Air Act, they must grant the United States the same rights. This stipulation causes difficulty in using Section 115 to control U.S.-Canadian acid rain pollution because of differences in the governmental levels where air pollution control authority is located. Canadian air pollution control authority is primarily vested in the provinces, while the U.S. Clean Air Act gives authority for control to the federal government. Although the Canadian government has amended its air pollution legislation to grant reciprocity to the United States, Section 115 has not been used successfully to control acid rain.[7] The Reagan administration has refused to allow the use of Section 115, since it considers the scientific evidence inadequate to support such action.[8]

Recommendations by the National Commission on Air Quality

The National Commission on Air Quality was established by the 1977 Clean Air Act amendments to review and analyze the effectiveness of the law in meeting its objectives and to "perform independent analyses of air pollution control and alternative strategies for achieving goals of the Act."[9] The commission plays an important role in preparing the ground for congressional action on reauthorization of the Clean Air Act. In *To Breathe Clean Air*, published in 1981, the commission discussed the problems associated with using the Clean Air Act to control acid rain. To overcome these problems the commission made the following recommendations:

1. Increase funding for carrying out acid precipitation research.
2. Increase funding for development of a long-term, nationwide atmospheric monitoring program.
3. Establish and implement a regional secondary standard for SO_2.
4. Enact a moratorium on relaxation of SIP SO_2 emission limits unless doing so will cause undue economic hardship.
5. Require a significant reduction in SO_2 emissions in the eastern United States by 1990.
6. Strengthen provisions requiring a state to reduce emissions that have adverse effects on other states.
7. Strengthen Section 126 to permit states to petition EPA for a finding that any aggregate of sources as well as any single source violates provisions against interstate pollution.

The first two recommendations—increased funding for research and monitoring—are designed to improve the scientific basis for controlling acid rain. Under current law, standards must be based on research findings. In addition, acid precipitation monitoring is an essential source of information for the development of sound methodologies for assessment and control programs. The third recommendation would control SO_2 emissions under a new regional secondary standard. Such a move would depart from the uniform national controls that operate under existing law and would address the regional nature of the problem. The fourth recommendation would force compliance with existing law by requiring sources to meet emissions limitations specified in State Implementation Plans. According to the commission's report, a 4–8 percent reduction in sulfur dioxide emissions could be achieved by enforcing provisions already contained in current law. In 1980 SIP relaxations resulted in a 1.5-million-ton increase in SO_2 emissions. If such relaxations continue, they would offset, by 1985, the anticipated decline in SO_2 emissions.

The commission's recommendation for significant SO_2 emissions reductions by 1990 would require extending strict New Source Performance Standards to existing stationary SO_2-emitting sources. Adoption of such a measure would go significantly beyond present law. Existing sources would be regulated for the first time under precise standards based on technological feasibility, an approach similar to that adopted by Sweden in the early 1970s.[10]

The commission recommended strengthening sections concerned with controlling transboundary air pollution. A state would thus be able to petition EPA to stop pollution from another state that caused adverse health and welfare effects. The language proposed by the commission removes the current requirement that the polluting source or sources must actually violate the receiving state's ambient air standards. The NCAQ recommendations also address the burden-of-proof issue. Under the proposed change it would no longer be necessary to identify an individual source causing pollution before EPA could intervene.

The recommendations by the National Commission on Air Quality, if adopted, would go a long way toward adapting the Clean Air Act to the new environmental situation created by acid rain. Congress has been considering the recommendations since 1981, and many amendments to the Clean Air Act have been inspired by the work of the commission.

Activities of Executive Agencies

Given the position of the Reagan administration on the economic burdens imposed on American industry by government regulation, it is not surprising that executive agencies have not taken a more active role in promoting acid rain control measures. The fundamental position of the administration is that the issue requires more study; and study it has received. Before late 1983 no executive agency had recommended control measures. And those recommended by the EPA administrator at that time were defeated at the cabinet level.

Title VII of the Energy and Security Act of 1980, the result of a proposal first made by President Carter in his second environmental message, established a ten-year program for the assessment of acid rain.[11] A National Acid Precipitation Assessment Plan was developed and the Interagency Task Force on Acid Rain was created to implement the law. The task force is housed in the President's Council on Environmental Quality and is jointly chaired by the Departments of Agriculture (USDA), Energy (DOE), and Interior (DOI); the Environmental Protection Agency; the National Oceanic and Atmospheric Administration (NOAA); and the Council on Environmental Quality. Senior officials of these six agencies form the Joint Chairs' Council. The task force also includes four presidential appointees and representatives of the national laboratories. Its main functions are to develop an acid rain assessment plan and to coordinate research efforts of federal agencies designed to develop a firm scientific basis for policy decisions. In 1985 the original ten task groups were reorganized into seven groups, with assessments being transferred to the Office of the Director of Research.[12]

The budget for acid rain research was steadily increased; between 1980 and 1983 the U.S. government spent $11 million on acid rain research. The president requested $27.5 million in 1984, and twice that amount for 1985. With supplemental funds granted by Congress, the total NAPAP budget was $33.1 million for 1984 and $65 million for 1985.[13] By 1986 it was $84 million, and in 1988 the budget was $85 million, of which EPA received $55.6 million. Table 7-1 shows the breakdown of expenditures by federal agency and program area for 1987.

During 1981 and 1982 the Reagan administration favored postponing action for controlling acid rain precursors until research guided by the Interagency Task Force was complete and the final report was submitted. The administration argued that any action taken now would be risky because controls would not ensure success in abating acid rain.[14]

In its annual report for 1982 the Interagency Task Force was more forthcoming, stating that "man-made atmospheric pollutants are probably the major contributors to acid deposition in Northeastern North America." This statement, according to the Interagency Task Force, is

Table 7-1.   National Acid Precipitation Assessment Program, Expenditures by Agencies, Fiscal Year 1987 (in millions of dollars)

Task Groups	EPA	USDA	NOAA	DOE	DOI	TVA	Total
I. Emissions and Controls	3.63			0.75			4.38
II. Atmospheric Chemistry	4.02		2.81	5.08		0.20	12.10
III. Atmospheric Transport	9.32		0.94	0.28			10.54
IV. Atmospheric Deposition and Air Quality Monitoring	5.42	0.68	0.43	0.81	1.88	0.04	9.26
V. Terrestrial Effects	9.86	9.99		0.04		0.40	20.29
VI. Aquatic Effects	20.01	1.40		0.25	2.37		24.03
VII. Effects on Materials and Cultural Resources	2.72				0.98		3.70
Assessments				1.85			1.85
Total	54.97	12.07	4.18	9.05	5.23	0.64	86.13

Source: National Acid Precipitation Assessment Program, *Annual Report, 1987* (Washington, D.C.: Government Printing Office, 1988), p. 9.

based on circumstantial evidence showing that the region of highest precipitation acidity coincides with the areas of greatest sulfur dioxide and nitrogen oxide emissions. The report states that the conclusion is also supported by meteorological studies tracking the physical movement of sulfur dioxide from sources of emissions to suspected areas of deposition. However, the report concludes current data and methods of assessment are "not sufficient to quantify the relationship between pollutant emissions and acid deposition on a regional scale or under varying conditions." It would also be impossible, given current information, to quantify the change in acid deposition that would result from a given change in precursor emissions. Statements about effects of acid rain are limited to damage to lakes in the Adirondacks. "Beyond the alteration of the chemistry and biology of certain sensitive surface waters, the other effects of acid deposition in North America are undetermined."[15] Assessment of possible damage to crops, buildings and bridges, and forests must await better documentation.

In sum, the Interagency Task Force abandoned the view that acid deposition in North America is primarily a natural phenomenon, but in other respects found the scientific evidence missing that would radically change the administration's position on acid rain.

Advocates of stronger environmental protection were gratified in 1983 when Anne Burford Gorsuch was replaced as head of the EPA by William Ruckelshaus. As the head of EPA when it was first formed during the Nixon administration, Ruckelshaus had gained the respect of environmentalists. During his confirmation hearings he promised the Senate Committee on Environment and Public Works that he would take up the acid rain question "with some degree of urgency."[16] Despite his espousement of a clear link between sulfur stack emissions and acid rain in July 1983,[17] he later told the House Energy and Commerce Subcommittee on Health and the Environment that it was premature to spend large sums of money on controls. His proposal for a small control program had been vetoed by the Office of Management and Budget.[18] In February 1988 it was still the official position of EPA that they were concerned about the acid rain problem but had insufficient knowledge to warrant control strategies.[19]

This view has been reinforced by the interim report of the Interagency Task Force.[20] Two years behind schedule, it was finally released in September 1987. The report consists of ten chapters that review scientific data relating to the physical processes and ecological consequences of acid rain, and the potential for emissions controls. An executive summary attempts to draw conclusions from the data presented and implies that it is premature to implement control measures. The report has been widely criticized both in the United States and Canada. The Canadians are particularly discouraged, claiming the conclusions often misrepresent the information contained in the main body of the report, contradict most of the previously published reports on the causes of acid precipitation, and are inconsistent with the spirit of the special envoys' joint report (see chapter 3).

The Federal/Provincial Research and Monitoring Coordinating Committee prepared a detailed critique of the report, which was submitted to Lee Thomas, EPA administrator, by Tom McMillan, minister of the environment, together with a letter in which he expressed his view that the report was "flawed, incomplete and misleading."

Congressional Response during the First Reagan Term

During the 1981−84 attempts to reauthorize the Clean Air Act, acid rain emerged as a primary issue. Four types of legislation were introduced that are relevant to the acid rain issue. They include bills to weaken the Clean Air Act ambient air standards (House only), research bills, bills to control interstate pollution, and bills to control acid rain precursor emissions. The bills to control interstate pollution would have allowed a state to show that it experienced adverse effects from another state's pollution without the necessity of proving that an individual source caused the problem. The other types of bills are discussed below.

Efforts to Weaken the Law

The major effort to weaken the Clean Air Act, supported by industry and the Reagan administration, was made during the 1981–82 House reauthorization attempt. The administration had initially sought a complete overhaul of the Clean Air Act, aiming to increase the role of the states and to advance the cause of deregulation. The power to set ambient air quality standards would have been returned to the states, and the federal government would have provided only guidance, primarily through research. This was the system that had been in effect prior to the 1970 Clean Air Act. The administration also attempted to require the use of cost-benefit analyses in establishing primary (health-based) standards. As time went by, however, the administration backed down and did not lend strong support to industry-backed legislative proposals, apparently under the influence of a 1981 Harris poll and in an attempt to avoid open conflict with Senator Robert Stafford, who was in charge of reauthorization in the Senate.[21]

The legislation introduced to weaken the Clean Air Act—primarily by allowing for less stringent SO_2 controls for stationary sources —was Representative James Broyhill's amendment, HR 3471, and Representative Thomas Luken's bill, HR 5252. The Broyhill amendment, with strong support from industry, would have eliminated the margin-of-safety requirement that is part of the current SO_2 primary standard, thus raising the standard from the current level of 80 micrograms per cubic meter to 105 micrograms per cubic meter. The latter number is the lowest exposure level at which human health effects of SO_2 exposure have been demonstrated. HR 3471 would also have given existing stationary sources an additional nine years to comply with primary air standards. The bill never received floor action. The Luken amendment contained many of the same provisions, but would not change the method by which ambient air quality standards were established. As a result, no increase in present emission levels for SO_2 would be allowed and current levels would remain unchanged until 1990.

With the exception of research proposals, acid rain amendments were not well received in the House Committee on Energy and Commerce markup sessions of the 1981 and 1982 reauthorization attempt. The committee never succeeded in reporting out a reauthorization bill because there was too much division among members over the kind of strategy to adopt. As the 1982 midterm elections approached, it became obvious that the majority of the House was not prepared to dismantle the Clean Air Act. A coalition in favor of strong air pollution regulation emerged, supporting amendments to control long-range sulfur dioxide pollution. In rejecting the decontrol attempt, the House took note of surprisingly strong voter support for the Clean Air Act. A 1981 Harris poll showed that 51 percent of the people surveyed wanted the law to remain unchanged, 29 percent wanted it tightened, and 17 percent wanted it relaxed.[22]

Control of Acid Rain Precursor Emissions

Support for strengthening the Clean Air Act and taking action against acid rain during the reauthorization attempts of the Ninety-seventh Congress first emerged in the Senate. Several amendments were introduced by members of the Committee on Environment and Public Works to limit environmental damage from acid rain, to be achieved by more stringent control of sulfur dioxide emissions from existing stationary sources.

A special acid rain control effort had originally been proposed by the National Clean Air Coalition, an organization comprising environmental, health, labor, and public interest groups.[23] The coalition proposed the following measures:

an emissions reduction program for acid rain precursors;

increased research aimed at identifying sources of acid rain;

a reduction in SO_2 emissions under a regional bubble approach (i.e., trading of emissions rights among sources in a given region) to encourage cost efficiency in meeting reduction targets;

prohibition of increases in sulfur dioxide emissions as a result of coal conversions (but sources would be allowed to use the bubble approach for their facilities);

provisions to prevent loss of jobs in the mining industry;

limitation on stack heights to reduce the long-range transport of pollutants.[24]

The proposed program attempted to combine reduction of acid rain precursor emissions with a softening of the economic impacts of new controls on the coal-mining industry.

The National Governors' Association endorsed the concept of an acid rain mitigation program that would reduce sulfur dioxide emissions on a regional basis. The governors recommended that a reduction target be set for the region, together with a three-year time limit to achieve the reductions. During this period the governors of the affected states would negotiate individual state reduction levels. The governors also proposed that an acid rain fund be established and used to support installation of scrubber control technology in order to meet the regional reduction requirements. The governors recognized that a new acid rain control program would tend to strengthen the federal role in regulation of air pollution. Their plan was designed to temper such an increase in federal authority and to retain state control over implementation of the program.[25]

Several of these proposals found their way into the Acid Rain Precursor Reduction Program that was developed by the Senate Committee on Environment and Public Works and included in Senate bill S 3041. The program had two major provisions—an emissions cap and a sulfur dioxide reduction requirement. The program would differ fundamentally from other parts of the Clean Air Act because regional rather than national controls would be used to control acid rain. The emissions cap would establish a thirty-one-state Acid Rain Mitigation region (ARM) to include all states east of the Mississippi and four states on the western Mississippi border. These states are located in areas where most of the sulfur pollution is emitted or damage from acid rain is evident. The emissions cap provides a short-run answer to acid rain by prohibiting relaxation of SIP requirements for specific sources of NO_x and SO_2 emissions. The cap would freeze emissions of sulfur and nitrogen oxides at 1980 levels. The freeze would be combined with a regional bubble concept under which new sources can buy emissions rights from existing sources that reduce emissions before obtaining a building permit. This provision is included to allow for continued economic growth in the affected areas.

The SO_2 emission reduction provisions of the ARM program are designed as a long-range solution to the acid rain problem. The required reduction was originally set at ten million tons and later reduced to eight million tons to be achieved within twelve years (1995), by which time all power plants would be controlled by New Source Performance Standards. This would represent a 35 percent decrease below 1980 emissions levels. The Senate bill included provisions for negotiations among states to achieve emissions reductions. The governors would have eighteen months to develop a plan, and when 75 percent of them were in agreement the plan would go into effect. Should the governors fail to reach agreement, EPA would set reduction levels for each state in proportion to their emissions levels in 1980. After state reduction requirements were decided, individual states would have to revise their state plans and submit them to EPA for approval. A state not complying with these provisions would be subject to controls equal to new source performance standards for coal-fired power plants. The feasibility of these provisions is questionable, however. States have experienced great difficulties in drawing up plans for allotment of emissions reductions among sources located within their territories. Negotiations involving thirty-one states would be much more difficult, since so many independent parties would be involved.

The Acid Rain Precursor Reduction Program does not specify the use of a particular technology in meeting reduction requirements. The bill, however, makes the following suggestions:

1. *Least-emissions dispatch*: using clean plants during light-capacity periods and dirtier plants only when needed during peak capacity operations.
2. *Early plant retirement*: building new, cleaner plants and shutting down older plants. To encourage this trend the law allows for the sale of emission credits supplied by retirement of older plants.
3. *Energy conservation*.
4. *Emissions trading and banking*: trading of emission credits between sources. This provides economic incentives because some sources can meet the standards more economically than others.
5. *Coal washing*: pretreatment of coal to remove sulfur.
6. *Fuel switching*: switching from high-sulfur to low-sulfur coal.
7. *Flue-gas desulfurization*: use of scrubber technology.[26]

The reactions to the acid rain provisions contained in S 3041 reflected the political positions of the commentators. In general, those supporting the bill argued that there was no time left to wait for conclusive research findings before implementing controls. Opponents argued that more research is needed to ensure that the desired results from sulfur dioxide reductions can be obtained before requiring industry to install expensive control technology. Senator Stafford would have preferred that his committee had approved a larger reduction, but felt that without this compromise "the alternative might have been no bill at all."[27]

Then-Senate Majority Leader Howard Baker cited two reasons for his opposition to the acid rain control program: economic growth would suffer, and there was no room for mid-course corrections should it later be found that utilities were not the main cause of acid deposition.[28] Senator Steve Symms opposed the acid rain program as the most expensive scientific experiment in history.[29] Similar comments were made by EPA officials, who argued that the program would unduly burden areas that use high-sulfur coal and result in significant increases in utility bills.[30]

The markup sessions in the Senate Committee on Environment and Public Works were not as controversial as those of the House committee, although there was disagreement between members on many issues.[31] By July 22, 1982, the committee had reached agreement on a reauthorization bill and reported S 3041. Although the Senate committee succeeded in reporting a bill during the Ninety-seventh Congress, none of its amendments reached the Senate floor, and they all died at the end of 1982. The debate over acid rain continued in the Ninety-eighth Congress. Many of the bills discussed in the previous section were reintroduced but died at the end of 1984.

The Second Reagan Term

With the passage of time the debate has become somewhat more focused. The question of scientific uncertainty has assumed a lesser role as an increasing number of people acknowledge at least a qualitative, if not a quantitative, relationship between industrial emissions and acid rain. The agenda is now dominated by economic and political issues. The economic issue is complicated by the fact that, as discussed in chapter 5, it is hard to quantify the costs of emission control programs, and even harder to quantify all of the benefits. Another aspect of this problem, which affects the politics of the issue, is regional differences. Without some kind of cost sharing, the states that produce and burn high-sulfur coal will continue to oppose controls.

Another facet of the political issue is the attitude of the Reagan administration. Its continued opposition to any kind of meaningful reduction in SO_2 and NO_x emissions has been one of the main impediments to passage of meaningful legislation. However, with the election in 1984 of a more conservative administration in Canada, there has been a slight moderation in the rhetoric coming from the White

House. In March 1986, following a summit meeting with Canadian Prime Minister Brian Mulroney, President Reagan acknowledged that "acid rain is a serious environmental problem in both the United States and Canada." While continuing to promote research initiatives rather than control measures, this apparent change in attitude led a significant number of moderate Republicans to support the concept of acid rain control.[32]

Clean air legislation received relatively little attention in 1985. However, 1986 saw a flurry of activity. The most significant bills were HR 4567 and S 2203. Introduced in April, with the support of 150 cosponsors, the Acid Deposition Control Act combines nationwide controls with a plan to share the financial burden. In 1984 acid rain legislation was defeated by one vote in the House Energy and Commerce's Subcommittee on Health and Environment. In 1986, with bipartisan support, it was referred to the full committee by a sixteen-to-nine vote. The overall objective of the legislation was to reduce SO_2 emissions by ten million tons annually and NO_x emissions by four million tons. Final reductions would be due by 1997, with an initial reduction required by 1993. Each state would be required to devise its own strategy for reducing emissions (the previous bill had mandated scrubbers on fifty of the largest sulfur polluters). An Acid Deposition Control Fund, financed by a nationwide fee on fossil-fuel electrical generation, would be used to limit local residential utility rate increases to 10 percent. NO_x emissions from cars would be reduced by imposing an emission limit of 0.7 grams of NO_x per mile by the 1989-model year (a standard already required in California).[33]

Despite widespread support for these bills, opposition from both the House and Senate leadership was instrumental in killing the legislation. Chairman of the House Energy and Commerce Committee John Dingell (D-Mich.) continued to oppose emissions controls for automobiles, and was able to prevent the bill from passing out of committee in time for a vote on the House floor.

By the end of 1986 there were 171 cosponsors in the House and 21 in the Senate. With momentum for acid rain controls building, January 1987 saw the introduction of two bills in the Senate. Senator Stafford's S 300, the New Clean Air Act, provided for an estimated reduction in SO_2 emissions of twelve million tons by limiting the total operating hours of major emitters of SO_2. Senator George Mitchell's S 321, the Acid Deposition Control Act, calls for reductions of SO_2 emissions by twelve million tons and NO_x emissions by four million tons. By October 1987 the Senate Environment and Public Works Committee (chaired by Senator Stafford) had passed the Clean Air Standard Attainment Act (S 1894) by a fourteen-to-two vote. Under Title II, which incorporates the provisions of S 321, it provides for a reduction in SO_2 emissions of twelve million tons by the year 2000.

However, there is still considerable opposition to the bill in the Senate as a whole. Majority Leader Robert Byrd (D-W.Va.) is worried about the effect of the bill on the coal industry in his home state. The continued opposition of EPA has also provided ammunition to opponents. In a minority report Senators Symms and Warner maintained that the costs for air pollution control under S 1894 exceed the benefits. Estimates developed by the Congressional Research Service and EPA indicate that the cost of air pollution control measures could increase from $31 billion annually to $70–$80 billion.[34]

HR 2666, introduced into the House in June 1987, is nearly identical to HR 4567. Hearings in Representative Waxman's Health and Environment Subcommittee began in February 1988, together with other air pollution legislation. Momentum for strong air pollution legislation continued to build in the House, where 193 representatives had signed a letter to the leadership of the House Energy and Commerce Committee urging prompt action on the Clean Air Act. A similar letter in the Senate had been signed by 42 members.[35] This sentiment was also evident in December 1987, when an amendment to extend compliance deadlines for cities exceeding ozone and carbon monoxide emissions levels for another two years was defeated by a vote of 257 to 162. By extending the deadline only eight months, Congress kept up the pressure to address air pollution issues. Otherwise, seventy urban areas will be subject to construction bans and withholding of federal highway funds.[36] To some extent the issue now is whether reauthorization of the Clean Air Act will result in an improvement in air quality, or whether weakening amendments will eliminate the opportunity to stipulate emissions reductions and control acid rain.

Research Bills

The Reagan administration has consistently claimed that it is premature to spend money on control strategies until more information on the causes and effects of acid deposition is available. One of the ideas promoted by the coal and electric industries is to invest in clean-coal technology rather than install emissions controls. Unfortunately, this technology is still several years away. In December 1985 the industry successfully lobbied Congress into providing $700 million for an Energy Department coal research program and $400 million over three years to pay one-half of the cost of building a commercial-sized clean-coal facility. With the endorsement of the special envoys' report (see chapter 3), President Reagan now wants a five-year, $5 billion government-industry program of research. With mounting budget deficits, Congress appears to have little appetite for such expenditures. In addition, the existing program is not being received with much enthusiasm by the utilities companies because the Department of Energy has stipulated that recipients of federal clean-coal dollars must repay the government if the projects they build are profitable.[37] Congressional action through March 1988 is summarized in table 7-2.

State Actions

In the absence of action at the federal level, a number of states have realized that initiatives at the state level may help to alleviate the problem of acid deposition within their borders. In general, less than one-half of the deposition is produced in-state, and some attempts at cooperative arrangements, particularly in New England, have been made. Although there is a diversity of approaches among states that have enacted legislation to control emissions, most have based emissions reduction targets on those anticipated in pending federal legislation. The following discussion illustrates the approaches taken, without offering a complete listing of states that have taken action.

Several New England states placed a cap on SO_2 emissions following a commitment made by their governors in June 1984. For instance, in Massachusetts a 1985 law required the Department of Environmental Quality Engineering to develop regulations in order to: (1) cap SO_2 emissions at 417,000 tons (the average for 1979–82, inclusive); and (2) reduce emissions if no national program has been enacted by 1987. Based on a national reduction standard of ten million tons per year, sources that burn more than one hundred million British thermal units of fuel per hour will be required to reduce SO_2 emissions to 1.2 pounds per BTU.[38]

Both New York and New Hampshire have programs that target sources emitting one hundred tons or more of SO_2. In New Hampshire an overall reduction of 25 percent is required by 1990, with a 50 percent reduction anticipated by 1995, based on a 1979–82 baseline. However, the implementation of the second reduction phase is dependent upon enactment of federal legislation.[39] A 1984 act in New York is based on a level of wet sulfate deposition of twenty kilograms per hectare per year. The preliminary Final Control Target, established at one pound of sulfur per million BTU for coal, and 1 percent sulfur by weight for oil, would meet New York's share of the national reduction

Table 7-2.   Synopsis of Proposed Acid Rain Legislation

Bill number	Date	Sponsor	Major provisions
		97th Congress	
S 1709	Fall 1981	Moynihan	Power plants that emitted more than 50,000 tons of SO_2 in 1980 required to reduce emissions by 85 percent. Allows trading of NO_x and SO_2 on a 2-to-1 basis.
S 1712	Oct. 1981	Dodd	Amends Section 110(a)(2)(E) and Section 126 to improve EPA's ability to take action against violation of interstate pollution.
S 2594	May 1982	Danforth	Reduces SO_2 by 7.5 million tons in 22-state ARM region. If research shows need, 9 states added to region and emissions reduced by an additional 2.5 million tons. Taxes utilities in the ARM region to pay for capital cost of controls.
S 2959	July 1982	Randolph	Accelerates research under Title VII of the Energy and Security Act of 1980 and caps SO_2 and NO_x emissions in ARM region.
S 3041	July 1982	Stafford	Reduces SO_2 emissions in ARM region by 8 million tons within twelve years. Allows governors to negotiate reduction requirements. Vehicle emissions standards for NO_x.
HR 3471	May 1981	Broyhill	Requires cost-benefit analysis in setting ambient air standards.
HR 4816	Oct. 1981	D'Amours	85 percent emissions reduction for fifty largest electric power plants in ARM region. Other plants generating more than 100 MW a year must meet 1.2 pounds of SO_2 per million BTU standard.
HR 4829/ HR 5555/ S 1706	Oct. 1981 Feb. 1982 Oct. 1981	Moffett/ Waxman/ Mitchell	Reduces SO_2 emissions in ARM region by 10 million tons within ten years.

Table 7-2.   (continued)

Bill number	Date	Sponsor	Major provisions
HR 4830/ S 3041	Oct. 1981/ Nov. 1982	Gregg/ Stafford	Research funding to accompany precursor control.
HR 5055/ S 2027	Nov. 1981/ Spring 1982	Rahall/ Byrd	Accelerates research under Title VII of the 1980 Energy and Security Act.
HR 5252	Dec. 1981	Luken	Relaxes automobile emission standards.
98th Congress			
S 454	March 1983	Byrd	Funds for accelerated research and mitigation of acid deposition effects.
S 766	March 1983	Randolph	Resembles S 2959.
S 768	March 1983	Stafford	Reduces SO_2 by 12 million tons in ARM region by 1998.
S 769	March 1983	Stafford	Resembles S 3041; amended to 10 million tons in committee.
S 877/ HR 4655	March 1983 Jan. 1984	Hollings/ D'Amours	Requires NOAA to report on wet and dry deposition with routine weather information.
S 2001	Oct. 1983	Duren- berger	Reduces SO_2 emissions in ARM region by 10 million tons. Taxes SO_2 and NO_x emissions.
S 2215	Jan. 1984	Glenn	Reduces SO_2 by 8 million tons in ARM region. Electricity tax subsidizes costs.
HR 3400	July 1983	Sikorski/ Waxman	Tax on electricity to subsidize installation of control technology. Reduces emissions from fifty largest sources in ARM region. Additional reductions nationwide.
HR 4404	Nov. 1983	D'Amours	Extends HR 3400 to reduce SO_2 emissions an additional 2 million tons; paid for by industrial sources.
HR 4483	Nov. 1983	Aspin	Taxes sulfur content of coal and provides subsidies for removal systems.
HR 4906	Feb. 1984	Rinaldo	Reduces SO_2 emissions by 10 million tons. Electricity tax subsidizes control systems.

Table 7-2.   (continued)

Bill number	Date	Sponsor	Major provisions
HR 5314	April 1984	Waxman	Incorporates provisions of HR 3400.
HR 5370	April 1984	Udall/ Cheney	SO_2 emissions reduced 11 million tons by 1996; polluters pay all costs.
HR 5590	May 1984	Green	Reduces SO_2 emissions by 10 million tons. Controls NO_x emissions.
HR 5592/ HR 5593	May 1984	Lloyd	Funding for clean-coal technology demonstration project.
HR 5970	June 1984	Vento	Reduces SO_2 emissions by 10 million tons. Fifty largest sources subject to emissions standards; utility generation fee to subsidize capital costs.

99th Congress

Bill number	Date	Sponsor	Major provisions
S 52/ S 283	Jan. 1985 Jan. 1985	Stafford/ Mitchell	Reduces SO_2 emissions by 10 million tons in ARM region.
S 503/ HR 2679/ HR 3677	Feb. 1985 June 1985 Nov. 1985	Proxmire/ Udall/ Solomon	Reduces SO_2 emissions by 10 million tons in ARM region in two phases. Accelerates research program.
S 1983	Dec. 1985	Kerry	Reduces SO_2 emissions by 12 million tons and NO_x by 3 million tons. Emission controls for high-sulfur coal users; utility generation fee to subsidize capital costs.
S 2203	Mar. 1986	Stafford	Reduces SO_2 emissions from industrial sources by 12 million tons by early 1990s. Reduces automobile emissions.
S 2813	Aug. 1986	Proxmire	Reduces SO_2 emissions by 10 million tons; full costs borne by utilities.
HR 1030	Feb. 1985	Conte	Resembles HR 5970.
HR 1162	Feb. 1985	Green	Reduces SO_2 emissions by 10 million tons. Controls NO_x emissions.

Table 7-2.   (continued)

Bill number	Date	Sponsor	Major provisions
HR 2631	May 1985	Whitley	Provides funds for research on effects of air pollution on forest decline.
HR 2918	June 1985	Rinaldo	Resembles HR 4906.
HR 4567	April 1986	Sikorski/ Conte/ Richardson/ Boehlert	Each state required to reduce acid rain pollutants. By 1997, SO_2 reduced by 10 million tons, NO_x by 4 million tons. Establishes an Acid Deposition Control Fund to limit residential utility rate hikes.
100th Congress			
S 95	Jan. 1987	Kerry	SO_2 emission reductions funded from a fee on emissions.
S 300	Jan. 1987	Stafford	Limits total operating hours for major emitters of SO_2. Reduces automobile emissions of NO_x and hydrocarbons.
S 316	Feb. 1987	Proxmire	Resembles S 2813.
S 321	Jan. 1987	Mitchell	Reduces SO_2 by 12 million tons and NO_x by 4 million tons. Depends on state programs to reduce overall emissions. Controls auto emissions. Provides funds for clean-coal technology demonstration projects.
S 1123	May 1987	Duren- berger	Reduces SO_2 by 12 million tons and NO_x by 4 million tons by 1997. Depends on state programs to reduce overall emissions.
S 1894	Nov. 1987	Mitchell	Contains provisions of S 321 as Title II.
HR 1664/ HR 1679	Mar. 1987	Solomon/ Cheney	Reduces SO_2 by 10 million tons in ARM region. Depends on state programs to reduce overall emissions. Provides funds for accelerated research program.
HR 2666	June 1987	Sikorski/ Conte	Resembles HR 4567.

Table 7-2.   (continued)

Bill number	Date	Sponsor	Major provisions
HR 2399	July 1987	De La Garza	Ten-year program to inventory forests for effects of air pollution on productivity.
HR 3054	July 1987	Waxman	Reduces SO_2 by 12 millon tons and NO_x by 4 million tons by 1996. Depends on state programs to reduce overall emissions.

target (assuming a ten-million-ton reduction). These controls would reduce emissions by 39 percent compared with 1980 levels. Legislative recommendations to meet these targets must be prepared by 1991. An Interim Control Target will achieve 40 percent of these reductions by 1990.[40]

Minnesota, recognizing the damage to its many lakes, passed legislation in 1982 to identify sensitive areas, adopt an acid deposition standard, establish a control plan, and ensure that all Minnesota sources are in compliance by 1990. With the most stringent standard yet adopted, Minnesota aims to protect its most sensitive lakes by restricting wet deposition to eleven kilograms per hectare per year in those areas. This target will necessitate a 24 percent reduction in emissions over 1980 levels, approximating that required in most national proposals. Because it proved difficult to establish source-receptor relationships, the Minnesota Pollution Control Agency will achieve this goal by placing a cap on emissions from its two largest sources, Minnesota Power and Northern States Power.[41]

Other states that have emissions control programs include Wisconsin, which requires a 50 percent reduction in SO_2 emissions by 1993 by switching to low-sulfur coal,[42] and Michigan, which has regulated SO_2 emissions since the early 1970s in order to meet federal ambient air quality standards.[43]

Conclusions

The threat posed by acid rain is not merely a threat against individual populations in neatly specified environmental settings. It is a subtle threat against the capacity of the natural environment to renew itself. Because acid rain causes subtle changes in the biosphere, the debate on measures to control acid rain has been slow to develop. Policymakers have seen no immediate threat to health and no need for urgency on this issue, even though acid rain may be a problem of global proportions. Because of the particular characteristics of acid rain, this problem will require a policy that is both comprehensive and geographically targeted. Uniform national standards will not address the problem efficiently because neither the causes nor the effects of acid rain are uniformly distributed across the country. It is clear, however, that efforts to mitigate local impacts of air pollution by dispersing pollutants more broadly have aggravated this problem. Certainly this situation suggests that a more efficient strategy in the long run is to stop pollution at its source. The questions again are: How can we do this? How much will it cost?

The burden that acid rain places on policymakers is the necessity of formulating a policy that is primarily concerned with the environment rather than with immediate health conditions. Health policy is certainly important, and the benefits to human health of past environmental legislation are certainly one important aspect of environmental policy. However, the rationale for environmental policy is that the natural environment has a value that is beyond our reckoning and is worth protecting to the extent possible for its own sake. Formulating an environmental policy will require policymakers to see beyond the immediate dramatic effects of pollution to possible subtle, long-term effects.

Formulating such a policy will require both energy and imagination, but these have not been lacking. Already, measures have been proposed which depart fundamentally from past regulatory patterns and which portend a more sophisticated and sensitive approach to environmental protection. Critics of acid rain control measures (and of environmental protection measures in general) point to the costs of past environmental legislation and the failures of past regulatory efforts as evidence that environmental legislation is too costly and too inefficient to be extended any further. The advocates of deregulation operate on the assumption that the unrestricted operation of the market or regulation at the state level can adequately address environmental concerns. But this has clearly not been the case. Past environmental policies may have been inefficient, but the process of imposing environmental regulations is being slowly refined. Critics fault environmental regulations for failing to solve problems in a few decades that industries took over a century to create.

The major obstacle to an acid rain control program has been economic. Policymakers have proposed measures that call for regionally targeted emissions reductions with considerable attention to the flexibility of these measures. Starting in 1983, there was an awareness that success depended upon more sophisticated financing measures that would mitigate the negative impacts of legislation on the coal industry and the Ohio valley.[44] These attempts have reduced the level of divisiveness based on regional differences. However, with a lack of leadership from the White House, there is still considerable reluctance on the part of many in Congress to spend money on an environmental program whose benefits cannot be accurately quantified.

||||||||

||||||||

Part Three
Supporting Structures

||||||||

Introduction to Part Three

The United States and Canada have a record of resolving environmental disputes successfully. Much of their success is due to an informal integration resulting from the shared language, culture, and economy of many of the people living on both sides of the border. These bonds have been strengthened since the beginning of the century by the establishment of institutions and procedures designed to address transboundary disputes of various kinds. By highlighting the strength and diversity of institutional relations between Canada and the United States, the following chapters show that a fuller use of existing structures is possible and that expanding the number and kinds of institutions concerned with acid rain can facilitate the development of an acid rain policy that will be acceptable to both countries. Chapter 8 uses the example of the International Joint Commission, and its role in the negotiations that led to the Great Lakes Water Quality Agreements of 1978, to make the point. Chapter 9 supplies an overview of a wide range of alternative policy mechanisms.

The 1972 and 1978 Great Lakes Water Quality Agreements are the most important examples of Canadian-U.S. environmental cooperation, documenting that the two countries can respond jointly to the problems posed by transboundary pollution. The agreements were facilitated by the existence of a mature, binational body (the International Joint Commission had been established in 1909) with experience in assembling policy-relevant information. The agreements did not settle all disputed points, but a process was initiated that allowed for continued cooperation between the parties.

The problems of water pollution in the Great Lakes and acid rain in North America are not directly comparable. First, the area affected by acid rain is much larger. Also, what was conceived in the past as a threat to a joint resource has become a polarized issue between polluting country and victimized country (though both sides realize that this is an oversimplification). But there are similarities. The water agreements were reached after national legislation had been enacted, and they proved helpful primarily in monitoring developments and coordinating national programs in their application across the border. Many years were needed to conclude the agreements. It took about eight years to agree on the terms of reference of a comprehensive assessment. A 1972 agreement allowed for further study and negotiations. In 1978 new deadlines and objectives were agreed to, including a mechanism for continued coordinated efforts to control pollution levels.

Another lesson from the work of the International Joint Commission concerns the relationship of the commission with governments. Both governments were reluctant to relinquish control to an international body. The commission was least effective when it tried to obtain new powers that would have gone beyond the limits set by the Boundary Waters Treaty. Any expectation that an international body for control of acid rain would actually implement policy is unrealistic. As in the case of the commission, the most useful role for an international body would be in monitoring emissions and deposition, and in evaluating control programs that have been agreed upon by the two governments.

Chapter 9 examines the role played in bilateral environmental matters in the past by the courts, by intervention in regulatory proceedings, and by relations among agencies and legislators. The analysis leads to a number of conclusions:

The use of joint committees and commissions such as the International Joint Commission have facilitated resolution of disputes involving other types of pollution.

Informal forums for policymakers on both sides of the border can contribute to the solution of transboundary pollution disputes by increasing their knowledge and understanding of one another's positions.

Existing cooperation between subfederal jurisdictions could be expanded to increase coordination of policies that affect transboundary environmental issues.

Formal channels ranging from lawsuits in foreign jurisdictions to intervention in regulatory proceedings are available to increase the attention given foreign concerns and impacts by both decision makers and the public.

A number of additional steps, some largely symbolic but still important, can be taken by both governments to demonstrate willingness to negotiate in good faith over acid rain. These range from equal-access reforms to changes in national statutes so that international impacts are recognized.

‖‖‖‖‖

‖‖‖‖‖

Chapter Eight
The International Joint Commission:
The Role It Might Play

‖‖‖‖‖

Introduction

In the negotiations on acid rain that have been under way since 1977, consideration has been given by the U.S. and Canadian governments to using the International Joint Commission (IJC) for monitoring the implementation of an air quality agreement.[1] In the past the two countries have made extensive use of joint committees and commissions, such as the Interparliamentary Group, the Committee on Joint Defense, and the International Boundary Commission, to resolve differences and promote intergovernmental communication.

This chapter asks whether it is feasible and appropriate to include the International Joint Commission, or some other binational organization, in the implementation of a Canadian–United States air quality accord. Analysis of the commission and its past efforts in bilateral control of air and water pollution suggests that the IJC has a useful role to play. Moreover, there is good reason to believe that it will be asked to become involved.

The IJC: An Overview

The International Joint Commission was created in pursuance of Article 7 of the Boundary Waters Treaty of 1909.[2] Its main functions, as defined in the treaty Preamble, are "to prevent disputes regarding the use of boundary waters and to settle all questions which are now pending between the United States and the Dominion of Canada involv-

This chapter was written by Paul Kinscherff and Pierce Homer.

ing the rights, obligations, or interests of either . . . and to make provision for the adjustment and settlement of all such questions as may hereafter arise."[3]

The IJC is composed of six members—three Americans and three Canadians. Canadian commissioners are appointed by the prime minister and U.S. commissioners are appointed by the president, with the advice and consent of the Senate. Decisions are based on a majority vote. Votes along national lines are rare; in over seventy years the commission has been deadlocked by a Canadian-American split only three times.[4]

The IJC maintains small headquarters staffs in both Ottawa and Washington, D.C., and a regional office in Windsor, Ontario. The headquarters staff assists the commission with planning, administrative, evaluative, government liaison, and public information activities. The regional office employs a few scientific experts, but for the last several years has primarily performed a public information function. The costs of the commission staff are divided equally between the United States and Canada.

The technical studies and fieldwork required by the commission's applications, reference, and surveillance/coordination functions described below are performed by binational advisory boards. Board members are usually well-qualified technical officials drawn from the civil service ranks of each federal government. Sometimes state and provincial officials also serve. Members are selected by the commission and are expected to operate on a collegial basis. Although their activities are financed by their respective governments, board personnel are expected to give their best professional advice to the commission, not simply relay their employers' respective viewpoints.[5] This has indeed been the case with the vast majority of advisory boards.

Initially, the IJC was concerned primarily with questions of boundary water levels and flows, navigation, and diversion. Its central mission was to ensure that boundary waters were shared equitably for the uses of commercial shipping, irrigation, and municipal water supply. After World War II, however, air and water pollution concerns assumed more importance on the IJC agenda. From 1909 to 1945 the IJC dealt with pollution issues only twice in fifty-two dockets; between 1946 and 1977 thirteen of fifty-seven dockets addressed either air or water pollution, or both. In one case the commission was asked to study and report on the "social problems of residents" as part of a boundary dispute in Point Roberts, Washington.[6] Such flexibility is characteristic of the governments' interpretion of the Boundary Waters Treaty and the uses to which the commission should be put.

The IJC has been used to find solutions to a number of bilateral problems involving water pollution, air pollution, water supply problems, navigation, power development, irrigation, recreation, and scenic beauty.[7] Under the Boundary Waters Treaty, it is responsible for:

approving or disapproving applications from the governments, companies, or individuals for the use, obstruction, or diversion of boundary waters;

investigating any differences arising between the two governments involving the rights, obligations, interests, and inhabitants of the other along the boundary;

monitoring compliance with the terms and conditions set forth in its approval of application and, when requested by the governments, monitoring or coordinating programs it has recommended;

arbitrating disputes between the governments.[8]

Thus, the commission exercises primary jurisdiction over the boundary waters that flow through the boundary of the United States and Canada; moreover, the IJC has the authority to require provision for the compensation of any interest along the boundary injured as a result of an approved application.

Arbitral powers are provided by Article 10, which states that the governments can submit disputes to the IJC for binding arbitration.[9] However, these powers have never been used. Several reasons for this have been offered by experts. First, there is a natural tendency on the part of both governments to settle disputes by negotiation and mutual agreement rather than by reference to a third party for binding arbitration. Second, on the U.S. side it would be extremely difficult to obtain the consent of the Senate. Third, there is a narrow view of the commission's competence in view of its identification with boundary waters, and a reluctance to strain commission resources by asking it to take on Article 10 functions. Finally, the time and costs involved in arbitrations are high.[10]

The IJC's investigative powers are provided by Article 9, which is the basis for the commission's important reference function. Under this article the commission is empowered to investigate "questions or matters of difference along the common frontier, called *references*, which are referred to the commission by the two governments. In such cases the commission reports the facts and circumstances to the governments of Canada and the United States and recommends appropriate action by them. The governments decide whether or not the commission's recommendations will be accepted or acted upon."[11]

This chapter is most concerned with the fact-finding role of the IJC. The governments can make references as broad or narrow in scope as they wish. The commission can neither initiate an inquiry on its own nor enforce compliance with the recommendations it formulates. Despite these limitations, the reference function has proven its value for promoting the settlement of bilateral disagreements.

Finally, the commission exercises administrative authority. At the request of the governments, it monitors or coordinates the implementation of its recommendations. This is generally done through follow-up references. The IJC also monitors compliance with its Orders of Approval, which spell out the conditions under which applications are granted.

The Reference Process

Initiation

The overriding consideration in any discussion of the IJC is that, except for the commission's primary jurisdiction over boundary waters, the "governments determine what approaches are to be used to settle boundary disputes or promote joint management schemes, and when and for what the IJC will be used."[12] Even though this does not give the IJC the right of initiative, the commission plays an important and vital role in boundary relations. For environmental relations the reference function is central. It is through the reference process that the commission provides requested information to the governments and proposes alternative courses of action to settle boundary disputes. Although not all references have dealt with pollution problems, all IJC pollution activities have come about through the reference process.

Although references can be submitted to the IJC unilaterally by either country, by tradition the two governments only submit questions on which they have come to prior agreement. In theory virtually any matter arising between the two countries could be the subject of a reference.[13] In practice the two countries often impose reference limitations which constrain the investigative authority of the commission and thus the impact that it can have on a given issue. For example, in a 1949 reference the governments requested that the IJC determine if the cities of Detroit and Windsor were being polluted by smoke, soot, fly ash, and other impurities, and to ascertain the extent to which vessels using the Detroit River were responsible. The terms of the reference limited the commission to making recommendations only on measures to reduce emissions from vessels plying the Detroit

River, thereby precluding any significant discussion in the IJC report about how to reduce pollution from nearby land-based industries.[14]

Results

The value of the reference process depends on the quality of IJC fact-finding and reporting and the ability and willingness of the governments to implement IJC recommendations.[15] Historically, the IJC has gained a reputation for reliable fact-finding. The commission is free to recruit top government scientists to serve on its boards, and IJC reports are credible discussions of the problems under investigation and the actions needed to rectify them.

Governmental ability and willingness to implement IJC recommendations depend on the nature of the recommendations and on bureaucratic and political obstacles. To be successful, IJC recommendations must be as specific as possible, emphasize short- and medium-term courses of action, and be politically acceptable to both the United States and Canada. In other words, the IJC must translate the findings of scientific research into meaningful action statements for nonscientist public policymakers. In general, the IJC has succeeded in providing the governments with useful recommendations in its reference reports, but it has experienced some difficulty securing governmental acceptance of its recommendations under the Great Lakes Water Quality Agreement. For example, some IJC recommendations have not received a formal response because they were vague and difficult to pinpoint.[16] Others were long-term, which made it difficult for the cognizant federal agencies to commit themselves to implementation because the responsible officials lacked the authority to do so. This reflects the fact that the level of IJC interaction with the governments, and especially the United States, fluctuates over time.[17] One possible cause for this is that governmental relations with the IJC have become routine, and consequently are delegated downward in the hierarchies of the U.S. Department of State and the Canadian Department of External Affairs. Another is that the relative importance and respect accorded the IJC varies, especially in the United States, from one administration to the next. Canada has historically viewed the IJC as more important and has been less motivated by domestic politics in appointing commissioners than the United States.[18] Confidence in the IJC fluctuates, and it was at an ebb in the mid-to-late 1970s.[19] The result was greater governmental resistance to the implementation of IJC recommendations.

Despite these considerable obstacles, the IJC remains an effective mechanism for alerting the governments to environmental and other problems along the U.S.-Canadian border. At present, government confidence in the commission is good.[20] In addition, as it has matured as an organization, the commission has become more adept at making politically acceptable recommendations and has developed the capacity to conduct references in a politically charged atmosphere. The following discussion outlines the role of the IJC in mediating disputes over water pollution in the Great Lakes and developing formal agreements between the two countries.

Use of the Reference Process for Water Pollution Control in the Great Lakes

The Great Lakes are the world's largest freshwater system, covering 84,000 square miles and draining an additional land area of 208,000 square miles. The lakes contain 95 percent of the U.S. fresh surface water supply. The Great Lakes basin spans 690 miles from north to south and 860 miles from east to west. It is home to 20 percent of the U.S. population and 25 percent of U.S. economic activity.[21] Fifty percent of Canada's population resides in the Great Lakes basin, while 60 percent of its economic activities operate there.[22] It has been home for over two hundred years to intense industrial and residential development. Great Lakes water has been used to dilute municipal wastes and cool nuclear reactors, and for drinking, recreation, fishing, and hazardous and industrial waste disposal.

The Boundary Waters Treaty of 1909 provides for the cooperative resolution of disputes concerning pollution of boundary waters. Article 4 of the treaty specifies that waters flowing across the boundary are not to be polluted on either side to the point of injuring human health or the property of the other country. Article 8 states that the two countries have "equal and similar" rights in the use of international waters. Article 4 implies that both countries are responsible for preventing Great Lakes pollution, while Article 8 implies that the two countries are entitled to an equal share of the lakes' assimilative capacity. Although the former view has generally prevailed in diplomatic discussions, it has been somewhat modified in practice. The Boundary Waters Treaty provides for an ongoing institutional mechanism to prevent and settle any disputes that may arise over the use of boundary waters. The IJC is a unique international body, whose composition and powers allow it to synthesize these two views of pollution in a pragmatic way. Its success as well as its limitations in facilitating U.S.-Canadian cooperation in cleaning up the Great Lakes give some indication of the organization's potential role in the future acid rain accord.

Responding to requests by the two governments between 1909 and 1964, the IJC developed a patterned response to the problems of boundary water pollution. In nearly every water pollution reference, the IJC made at least two recommendations: the U.S. and Canada should adopt mutual and compatible water quality objectives; and the two countries should appoint bilateral advisory boards to monitor and report on progress in meeting those objectives. Neither recommendation lends itself to easy implementation. The adoption of "mutual and compatible" water quality objectives is a difficult task for both technical and political reasons, and continuing surveillance over water quality by the ad hoc advisory boards is not a responsibility specifically conferred upon the IJC by the Boundary Waters Treaty.[23] Yet the existence and activities of these boards have been indispensable to U.S.-Canadian water pollution control efforts.

The first IJC water pollution reference came in 1912. Prompted by typhoid outbreaks in Detroit and other areas, Canada and the United States asked the IJC to investigate water pollution in the Detroit and Niagara rivers. In 1918 the IJC issued its report. It confirmed the existence of pollution in the connecting channels of the Detroit and Niagara rivers, and found that this pollution was in violation of Article 4 of the Boundary Waters Treaty. To resolve the issue, and similar ones in the future, the IJC proposed a strong jurisdictional role for itself.[24]

Negotiations on these recommendations continued intermittently until 1929, when they were terminated by the United States, which was unwilling to grant the power of final determination of fact to the IJC. Moreover, the original reason for the reference—the existence of typhoid—disappeared with the introduction of chlorine to municipal water supplies. Canada proposed reopening these negotiations in 1942, but the United States refused, calling the timing "inappropriate."[25]

The second IJC water pollution reference did not occur until 1946, when the two governments directed the IJC to investigate pollution in the connecting channels of the St. Clair River, Lake St. Clair, and the Detroit and St. Marys rivers. The Niagara River was added to the reference in 1948. In 1950 the IJC reported that each channel was seriously polluted, and made two more modest proposals to the United States and Canada. First, it reiterated its position that the governments should adopt mutual objectives for boundary water quality control; second, it recommended the establishment of two separate advisory boards for the Superior-Huron-Erie and the Erie-Ontario sections of the channels.[26] Canada and the United States quickly accepted these proposals, establishing the model for future water pollution references to the IJC.

Some observers see the adoption of mutual water quality objectives and the appointment of bilateral advisory boards, recommended in three succeeding references by the IJC,[27] as weak substitutes for the IJC's lack of enforcement powers.[28] What has evolved since 1909, however, is an effective set of cooperative mechanisms that respect the different governmental procedures in the United States and Canada, and reflect the multijurisdictional nature of boundary water pollution in the two countries. The appointment of bilateral investigatory and monitoring boards has grown into the most dynamic element of the IJC formula for responding to boundary water pollution. The boards ensure the representativeness of the process by including, at a very early stage of the reference process, every affected state and federal agency. The composition and structure of the IJC boards have built partisan representation into a process often characterized as collegial and impartial. However, one Canadian official has called the process "an international monument to the effectiveness of Federalism."[29] Another has said that "the members of such a board cannot completely divorce themselves from their national origins so that even in the production of the IJC Report there was an element of give and take, of compromise—I won't say negotiation—in order to arrive at agreed findings."[30]

The Great Lakes Water Quality Agreements

The Role of the IJC

The Great Lakes Water Quality Agreement of 1972 began with a 1964 reference to the IJC on pollution in Lake Erie, Lake Ontario, and the St. Lawrence River. The impetus for this reference came from the scientific, diplomatic, and policy communities, and preceded the heightened public concern over water pollution that became evident in the mid-to-late sixties.

The IJC followed its established practice of appointing bilateral advisory boards to investigate pollution of boundary waters. Many of the persons appointed to the lower lakes investigatory boards held positions in agencies that would later be involved in pollution abatement in the Great Lakes basin.[31] Many of the advisers to the IJC, in other words, were making recommendations they would later have to implement on state and federal levels.

The investigatory boards delivered a final report to the IJC in the fall of 1969. The commissioners were pleased with the end product and promptly made public the most detailed and complex hydrological study of its time. In addressing the issues raised in 1964, it concluded that the lower Great Lakes were seriously polluted on both sides of the boundary to the detriment of both countries, a violation of Article 4 of the Boundary Waters Treaty. Without identifying specific sources of pollution, it established that the principal causes were wastes discharged by municipalities and industries into the boundary waters and their tributaries, primarily on the U.S. side. Both lakes were in advanced stages of eutrophication, and a controlling factor in this process was the discharge of phosphorus from detergents and other municipal and industrial wastes. Urgent measures were needed, and these were spelled out in detail by the commission.[32]

In responding to the 1969 lower lakes report, the IJC followed its established practice of calling for the adoption of common water quality objectives and the establishment of bilateral advisory boards to monitor the achievement of those objectives. The IJC also called for a timetable for the construction of pollution abatement facilities, effluent objectives (in addition to water quality objectives, which specify the desired condition of the receiving waters), and the establishment of a regional office to provide enhanced technical support and coordination for the IJC.[33]

Recognizing a political opportunity following a heated confrontation over detergent phosphorus at a public hearing, the IJC decided not to wait until the fall to issue its final report. It forwarded a special interim report to the two governments in April 1970, identifying phosphorus as the most prominent and yet most readily controllable pollutant in the lower lakes. The IJC report stated that "a reduction in the phosphorus inputs could be accomplished most quickly by limiting the phosphorus content of household detergents."[34] Canada responded to the recommendation and in August 1970 enacted the Canada Water Act, which immediately limited the phosphate content of detergents to 20 percent, to be further reduced to 5 percent by 1973. The United States appeared to be heading in the same direction.

The Negotiations

The late sixties and early seventies were landmark years for environmental legislation in both the United States and Canada. President Nixon signed into law extensive air and water pollution control acts and created the Environmental Protection Agency (EPA). The national media focused attention on environmental problems and specifically posed the question of whether Lake Erie was dead or dying. In the early 1970s it was not politically possible to be against programs to control water pollution. Concern over Great Lakes water quality may have originated in the scientific and policy communities, but it flourished in the political climate of the day.

The possibility of a bilateral solution to water quality problems in the Great Lakes became an attractive proposition to a number of Nixon appointees. Daniel Moynihan of the White House, Russell Train of the newly created Council on Environmental Quality (CEQ), the State Department's Office of Environmental Affairs, and the Federal Water Quality Administration all viewed a cooperative agreement on the Great Lakes as a valuable piece of bureaucratic turf. These agencies and personalities all competed with one another for control over a Great Lakes initiative, whatever shape it might take. Canada meanwhile pushed for some kind of joint initiative. Canadian efforts were rewarded when Train and the CEQ were charged with drafting a response to the IJC report and initiating talks with the Canadians.

Formal negotiations on Great Lakes water quality took place from May 1970 to April 1972.[35] Preliminary meetings in May and June of 1970 proved the value of an established framework for addressing Great Lakes pollution. With the preliminary results of the IJC reference in hand, the two governments were able to agree that transboundary water pollution did exist, that the United States was responsible for a majority of it, and that the conditions were contrary to the obligations of the Boundary Waters Treaty. The United States did not accept IJC interim recommendations to limit phosphorus inputs, but it did agree to the establishment of a joint working group that could expand and articulate the areas of agreement and delineate the areas of disagreement between the two countries.

The joint working group functioned very effectively from September 1970 to April 1971. It was divided into ten subgroups, and within each subgroup the interests of the two federal governments, Ontario, and the eight Great Lakes states were represented. This composition and structure allowed for the development of a difficult and complex intergovernmental relationship in an informal and highly productive setting. In addition, Russell Train's September 1970 report on Great Lakes water quality gave direction and impetus to the working group proceedings by reaffirming some of the IJC's interim recommendations. By the fall of 1971 an agreement appeared imminent.

However, other developments in the United States and Canada stalled diplomatic negotiations. The Nixon administration suddenly reversed its course on the phosphates issue, claiming that the leading phosphate substitute might exert harmful effects on human health. Despite a lack of scientific evidence, the administration advised the public to return to the use of phosphate detergents. An immediate uproar ensued, proposals for U.S. phosphate controls were aborted, and bilateral talks—in progress since June of 1970—were stalled for months. The phosphate issue was never satisfactorily resolved, and Ottawa had to provide Ontario with funding for an accelerated program of municipal wastewater plant construction. Both events illustrate the multijurisdictional nature of the problem and the sometimes untraditional alignments of interest that can occur in such a setting.

In December 1971, however, negotiations resumed and the two countries compared draft texts of proposed agreements. The working group had successfully narrowed the areas of disagreement to four: equal rights to the assimilative capacities of the lakes; timetables for the attainment of common water quality objectives; new responsibilities for the IJC and an IJC regional office; and phosphate reductions. It is significant that none of these four issues involved a substantial measure of scientific disagreement between the two countries. Between January and March of 1972 the Canadians yielded their positions on all four areas of disagreement. Of critical importance was Canada's decision not to insist on the right to 50 percent of the assimilative capacity of the lakes. U.S. officials contended that Article 4 of the Boundary Waters Treaty prohibits any pollution of the Boundary Waters and so obligates both sides to control transboundary pollution. The Canadians argued that since Article 8 of the Boundary Waters Treaty guarantees "equal and similar" use of the boundary waters, each side has an effective right to contribute pollution up to 50 percent of the "assimilative capacity" of the waters. Acceptance of the Canadian position would have required drastic U.S. control measures and a minimal cutback in Canadian water pollution in the Great Lakes.[36] The U.S. position was that, when pollution exists in the Great Lakes, both nations have an obligation to correct it.[37] This ultimately became the basis for the Great Lakes agreements.

Canadian efforts then focused on securing firm U.S. commitments and timetables for the attainment of common water quality objectives. The U.S. side was under heavy pressures to keep expenditures down. Canada reluctantly accepted the wording that ultimately became Article 5 of the Great Lakes agreement: "Programs and other measures directed toward the achievement of water quality objectives shall be developed and implemented as soon as practicable in accordance with legislation in the two countries. Unless otherwise agreed, such programs and other measures shall be either complete or in the process of implementation by December 31, 1975."[38]

This final phrase of Article 5, with its deadline, became more important in later years when U.S. municipal treatment programs lagged behind schedule, federal wastewater funds were impounded, and sewage treatment costs exceeded earlier estimates.

Two issues remained. The State Department opposed the establishment of an IJC regional office, and a schedule of reductions for U.S. phosphate inputs had to be agreed on. Christian Herter, Jr., U.S. IJC section chairman, broke the impasse and obtained congressional support for the funding of a small regional office. Canada agreed to a provision that authorized but did not require the establishment of a regional office. In March 1972 Canada also agreed to an American proposal for scheduling phosphate reductions, even though the reductions were less ambitious than the ones proposed by the IJC in 1970.[39]

The 1972 Agreement

On April 15, 1972, Prime Minister Trudeau and President Nixon signed the Great Lakes Water Quality Agreement. The agreement sketched out an unusual project in international cooperation, the status of which was to be assessed in five years. In sum, the agreement:

established general and specific water quality objectives for the lakes;

designated December 31, 1975, as the date when programs to achieve those objectives were to be completed or in progress;

assigned new responsibilities and functions to the IJC to implement the agreement (including opening a regional office and the right to publicize its findings);

authorized the establishment of two international boards—the Water Quality and Research Advisory boards—to oversee and implement the agreement.[40]

Equally important were the subjects not addressed by the 1972 agreement. By authorizing a Research Advisory Board, the United States and Canada recognized the high degree of scientific uncertainty still attached to water pollution in general and to the large, geologically complex Great Lakes in particular. The agreement also left for future study the problems of vessel wastes, pollution of Lakes Superior and Huron, nonpoint sources of pollution, industrial pollutants, toxic substances, radioactivity, and thermal pollution.

Although the emphasis of the 1972 agreement was clearly on municipal and industrial point sources of pollution, particularly sources of phosphorus, the United States and Canada recognized that the joint management of a complex and seriously polluted body of water like the Great Lakes would require a good deal of flexibility. Priorities and objectives might change over time. Research might indicate that certain types of pollution control are ineffective, or that certain classes of pollutants pose a larger threat than previously imagined. One kind of objective might be reached and bilateral resources might need to be reallocated. The 1972 agreement allows for a flexible and progressive set of responses to water pollution. It was not so much an agreement set in stone as the initiation of a process of pollution control that required coordinated efforts on both sides of the Great Lakes. The 1978 agreement continued and expanded that process.

The 1978 Agreement

At the end of the first five years it was apparent that many of the objectives contained in the 1972 agreement could not be met and that a general revision of the original objectives was in order. The Great Lakes Water Quality Agreement of 1978 set new deadlines for municipal wastewater programs, revised the general and specific objectives contained in the 1972 agreement, redefined the role of the IJC regional office, and adopted a basinwide or "ecosystem" approach to the problems of water pollution. Developed by the Great Lakes Advisory Board, the ecosystem approach to water quality recognizes that water quality problems originate from a variety of human activities. A comprehensive agreement on Great Lakes water quality has to address these complex interactions, allowing for modifications as scientific and technical knowledge evolves.

The purpose of the 1978 agreement was "to restore and maintain the chemical, physical, and biological integrity of the waters of the Great Lakes Basin Ecosystem."[41] The United States and Canada therefore agreed to develop and implement programs and measures to control phosphorus inputs, to eliminate toxic discharges, and to minimize hazardous wastes in the Great Lakes. These three commitments were not only more detailed than those laid out in the 1972 agreement, they were also subject to future revisions on the basis of programs of research and monitoring. Allowing for incorporation of the results of future IJC research and monitoring, the agreement provides that the specific objectives and annexes can be amended at any time by the mutual consent of both parties.[42]

The move from the 1972 to the 1978 agreement reflects better knowledge about water pollution in general and pollution in the Great Lakes in particular. The 1978 agreement captured this dynamic; comprehensive environmental policies must be based as much on scientific questions as on scientific certainty. Water quality policies in the Great Lakes must be adaptable to developing scientific and technical knowledge.

Lessons for Acid Rain Control

Prior to acid rain, pollution in the Great Lakes was the most complex bilateral issue in environmental policy faced by the United States and Canada. The policy response that had to be developed was not of continental scope, but even so, controls required the coordination of two countries, eight states, one province, and hundreds of municipalities. There are many parallels and shared themes between control of pollution in the Great Lakes and control of acid rain. However, the Great Lakes Water Quality Agreements of 1972 and 1978 do not chart a path for resolving the acid rain conflict. The geographic area covered by the Great Lakes agreements is large, but it is dwarfed by the region that will have to be considered to control transboundary air pollution. Despite success in cleaning up the Great Lakes, neither Canada nor the United States has technical, institutional, or political experience in dealing with a problem of this magnitude.

The design of the 1972 Great Lakes Water Quality Agreement, although far from textbook perfect, offers a good example of self-conscious, bilateral cooperation. For that reason alone its lessons may be limited to truly cooperative situations. So far, acid rain does not fall into this category. But it may in the future, if evidence and perception of damage become more equally divided between Canada and the United States. This analysis of the Great Lakes Water Quality Agreements has identified nine themes that provide some insight into the current discussions concerning acid rain:

1. The threat to Great Lakes water quality involved many jurisdictions and interests, and not just two national governments.
2. The United States and Canada disagreed on whether they had an "equal right" to pollute or an "equal obligation" not to pollute the Great Lakes.
3. There was a limit to collegial and "impartial" decision making. Injecting an element of partisan representation into the working groups and technical boards facilitated, rather than hindered, progress toward an agreement.

4. Interest in Great Lakes water quality first surfaced in the scientific and policy communities. The general interest in environmental policy came later.
5. There was broad popular support for environmental initiatives at the time of the first Great Lakes agreement.
6. The United States found it difficult, if not impossible, to target environmental policies to the specific needs of the Great Lakes basin.
7. Even though U.S. positions generally prevailed in the 1970–72 negotiations, the United States and Canada treated Great Lakes water quality as a *common problem*. Also, the working group was able to limit the scope of negotiations to four comparatively narrow, nonscientific areas of disagreement.
8. The 1972 agreement made no attempt to be comprehensive; it focused on the most pressing issues of the day and left others for the five-year review; and it did not create any new domestic programs for pollution control in the Great Lakes.
9. The 1978 agreement was based on the recognition that "scientific certainty" is not a static concept. Scientific and technical knowledge continuously evolve and policies based on them must take this into account.

The water quality agreements did not create any new programs for water pollution control. They relied instead on existing or proposed domestic legislation and provided a focus for their coordination. There are no current U.S. or Canadian programs to control SO_2 or NO_x emissions from all existing point sources. A bilateral effort to control acid rain will have to be based on new domestic legislation. After these measures are in place, or at least in the offing, the kinds of coordinating activities agreed upon in the Great Lakes agreements become possible.

In 1972, and to a large extent in 1978, there was no organized opposition to the Great Lakes agreements. It was difficult to be "against clean water" and the variety of affected uses—industry, municipal wastes, forestry, agriculture, and shipping—virtually precluded the organization of an effective opposition. In the acid rain debate, however, a small group of energy and industrial activities *are* affected. They have well-organized trade groups and very liquid political assets. Theirs is an almost "natural" organization held together by a robust group of political action committees.

The Great Lakes Water Quality Agreement of 1972 was signed eight years after the 1964 IJC lower lakes reference (which itself took eight years [1956–64]). A total of sixteen years was needed to produce a comprehensive, mutually acceptable assessment of Great Lakes pollution. On this continent attention was first called to the problems of acidic depositions in 1976. Although it is not a new policy issue, acid precipitation in North America is perhaps twelve years old, and may require additional time before the many strands cohere into a single thread of policy. In particular, the lack of comprehensive, mutually acceptable assessments of acidic deposition in North America has prevented the two countries from sketching out broad areas of agreement and disagreement. Policy development on acid rain has not met certain expectations, yet it is not slow by comparison to the Great Lakes agreements.

Issues of Great Lakes water quality were perceived with greater urgency than those having to do with acid rain for three reasons. First, the health effects of acid rain are subtle, long-term, and speculative. Long-term risks are more difficult to evaluate than immediate threats from contaminated water and food supplies. In 1972 people saw very direct health consequences from a failure to act on Great Lakes water quality. The same is not true in the case of acid rain. Second, market surrogates had been established for many of the uses of Great Lakes waters, and dollar values could be placed on many of the projected losses. The same is not true in the current acid rain discussions. Third, and perhaps most important, acid depositions and their long-range transport are more difficult to understand than transboundary water pollution. The electricity consumer in Ohio may not see or care about the connection between his water heater and the possibility of aluminum traces in Canada's water supply. The relation between household detergents and algae blooms on Lake Erie is not only "in his back yard," it is conceptually easier to grasp. The transition between scientific explanation and policy prescription may have been easier in the case of Great Lakes water quality than it is in the case of acid rain.

Air Pollution References

IJC activities have historically focused on transboundary water pollution and flows. This is not surprising. The Boundary Waters Treaty gives the IJC primary jurisdiction over transboundary water management and specifically prohibits boundary water pollution which would harm the health or property of either side. But because the treaty does not restrict the uses to which the IJC can be put, and because it has become increasingly apparent that boundary water degradation is in part caused by atmospheric pollutants, the governments have intermittently referred questions concerning air pollution to the IJC.

The commission first became involved with problems of air pollution in 1928, when it was asked to investigate and report on the extent of damage in the state of Washington caused by fumes from a smelter at Trail, British Columbia.[43] In this now-famous dispute, the IJC recommended a lump-sum payment of $350,000 to cover all damages through 1931 and recommended remedial measures to reduce further emissions from smelter fumes. The case was not settled until after 1941 (by an independent tribunal) because the United States was unwilling to accept the commission's recommendations.[44]

Detroit-Windsor

Air pollution in the Detroit-Windsor area has been the subject of periodic commission investigations since 1949, when the governments requested an IJC report on the problem of smoke from area vessels plying the Detroit River. As noted earlier, the terms of the 1949 reference were somewhat constraining. In 1960 the IJC reported that, although vessels did contribute to the area pollution problem, various industrial, domestic, and transportation activities were mostly responsible.

Six years later the governments requested that the commission determine if the ambient air quality of the Detroit-Windsor and the Port Huron—Sarnia areas were polluted to the detriment of health and property. In a significant move, the governments also asked the commission to take note of air pollution problems in all other boundary areas and, when appropriate, draw the problem to the attention of both governments. To fulfill the terms of this reference the commission established the International St. Clair—Detroit River Areas Air Pollution Board and the International Air Pollution Advisory Board.

The International St. Clair—Detroit River Areas Air Pollution Board completed its investigation and submitted a detailed report to the commission in 1971 that confirmed the existence of significant regional transboundary air pollution. In 1972 the commission forwarded its report to the governments. In addition to reporting on the nature and extent of the regional air pollution problems, the IJC recommended specific boundary area air pollution objectives for suspended particulates and sulfur dioxide.[45] The commission also urged that the state of Michigan and the province of Ontario cooperate to accelerate existing abatement schedules and to prevent the creation of new pollution sources.[46]

In 1974 Michigan and Ontario signed a memorandum of understanding (MOU) pledging such cooperation through their existing Michigan-Ontario Transboundary Air Pollution Committee. The MOU established December 31, 1978, as the target date for achieving the air pollution objectives recommended by the IJC. Michigan and Ontario also requested that the IJC assume responsibility for the monitoring and implementation of air pollution control programs in the Detroit-Windsor and the Port Huron—Sarnia areas.[47] As a result of a 1975 follow-up reference, the International Michigan-Ontario Air Pollution Board was established to assist the commission by coordinating surveillance and monitoring programs related to air quality in the region. The commission reports annually to the governments on the results of its activities.

The International Air Pollution Advisory Board

The International Air Pollution Advisory Board was created under the 1966 reference to assist the commission in its efforts to alert the governments to transboundary air pollution problems. Because its mandate is somewhat open-ended, the board has been involved in a variety of activities. For example, in 1974 emissions from a Boise-Cascade Kraft mill in International Falls, Minnesota, increased suddenly as a result of control equipment failure. The increased emissions presented a possible health hazard to residents of Fort Francis, Ontario. The Ontario Ministry of the Environment requested the urgent assistance of the IJC in persuading Minnesota to take action. The board recommended that the commission take up the issue with the U.S. and Canadian governments, and additionally advised that the Minnesota State Pollution Control Agency be urgently requested to take immediate action.[48]

In another instance the board requested the Federal Energy Administration to give prompt and favorable consideration to an application submitted by the city of Detroit for allocation of fuel oil as part of a local program to reduce particulate emissions from a city-operated generating plant.[49] The board has also investigated the possible adverse effects of proposed power plants along both sides of the border.

During the 1970s some board activities led to expressions of concern by the governments. A 1974 board report had raised questions about adverse effects of fluoride emissions from a Reynolds Aluminum plant near Massena, New York, on the St. Regis Indian Reserve cattle industry. In 1977 the board held an open meeting of interested persons, and its report concluded that this pollution problem needed bilateral as well as domestic remedial action. The governments, however, felt that the mandate of the 1966 reference had been exceeded. The scope of the board study, including a public meeting, was seen as going beyond the IJC's alerting role, and there was some perception that the commission also aspired to a role in solving the problem.[50] The incident resulted in an October 1978 commission letter to the board limiting its activities to alerting rather than in-depth study and investigation.[51]

Summary

The experience of the commission under the 1949 and the 1966 air pollution references further illustrates the two points raised earlier. First, if the IJC is careful to restrict its fact-finding activities to the questions specifically brought up by reference, its reports stand a good chance of receiving favorable governmental consideration. Once accepted by the governments, commission reports can provide the basis for bilateral action. As a result of the commission's investigation of air pollution caused by vessels plying the Detroit River, a subsequent general air pollution investigation was requested by the governments. Second, whenever commission activities stray beyond relevant reference limitations, swift and forceful governmental objections invariably follow. The IJC can act as a monitor and coordinator of governmental activities, and has done so on several occasions; however, the commission acts at great risk when it assumes the role of problem solver without prior government approval.

Responses to Acid Rain

Acid rain emerged as a policy issue after scientific research into the basic nature of the problem became available.[52] Although North American attention was originally a result of European research, interest broadened when IJC studies on pollution of the upper Great Lakes found that a surprisingly high proportion of pollutants entered from the atmosphere.[53] It was through these studies, conducted under the authority of the 1972 and 1978 Great Lakes Water Quality Agreements, that the IJC became the first bilateral institution (although not the first North American organization) to address the problem of long-range transport of air pollutants. The commission, along with its science advisory board, originally stood alone in publicizing concern over the issue.[54]

The IJC first expressed concern over atmospheric pollution of the

Great Lakes in its 1974 *Third Annual Report on Great Lakes Water
Quality*. The report proposed a Water Quality Surveillance Program
that included an atmospheric fallout program for Lake Erie.[55] It was
not until 1976, however, that the long-range transport of air pollutants
was reported by the commission as a potential problem. In March of
that year the commission brought to governmental attention the need
for a common boundary study on the problem. The Great Lakes Water
Quality Board pointed out the need for a regular surveillance program on airborne pollutants in its 1976 annual report. The recommendation was based on estimates by the Upper Lakes Reference
Group that atmospheric deposition accounted for as much as 30–40
percent of the total copper and lead loadings in the upper Great Lakes.
Canada responded to the commission's concerns by agreeing with the
need for a joint study, and noted that one would be undertaken at the
earliest date convenient to the U.S. authorities.[56]

In March 1977 the International Air Pollution Advisory Board recommended the establishment of a long-range air pollutant transport
research group structured as a separate commission board.[57] The
commission acted upon this suggestion by advising the governments
of the apparently significant pollution of the Great Lakes from atmospheric fallout and the need to establish a mechanism for achieving
international coordination and cooperation on research related to
long-range transport of airborne pollutants. Later that year a commission follow-up letter expressed the need for clear principles of equita-
ble utilization of the atmosphere and international obligations, including prior consultation as well as objectives and machinery to monitor
compliance. The commission felt the need to deal with international
air pollution on a continuing basis.[58]

In November 1978 the governments responded to the IJC recommendations by informing the commission of the establishment of the
Bilateral Research Consultation Group (BRCG). Although board members could also be members of the group, they would serve as government officials and not as IJC representatives. The BRCG was thus
created as an entity of the governments, independent of the IJC.[59]

One reason the IJC was excluded from the Bilateral Research Consultation Group may have been that at the time the relationship
between the commission and the governments was strained. Two
recent commission letters had been badly received. The first ques-
tioned whether the governments had the right to decide, by themselves, if matters fell outside commission jurisdiction. The other was
considered highly presumptuous by some officials because it suggested that the IJC might be used in some way for notice and consulta-

tion purposes between the governments. As a result, U.S. and Canadian officials adopted a tougher, more distrustful attitude toward the commission, and the governments may have been reluctant to include the IJC in new bilateral initiatives.[60] Another reason for not turning to the IJC may have been its regional focus on the Great Lakes area. Long-range transport of air pollutants is important for the Great Lakes, but not acid rain as a particular form of long-range transport. Because of their large size, the lakes by themselves are not considered to be sensitive to acid rain. But the same is not true, obviously, for some of their watersheds.

Even so, IJC boards continued to expand the base of scientific knowledge about long-range transport of airborne pollutants in 1978. That year both the Upper Lakes Reference Group and the International Reference Group on Land Use Activities capped six years of investigation by submitting their final reports under the Great Lakes Water Quality Agreement of 1972. These reports reiterated past findings that atmospheric sources contribute large amounts of pollution to the Great Lakes, both from direct fallout into the Great Lakes and from land runoff into the Great Lakes basin.[61] The commission noted these problems in its *Sixth Annual Report on Great Lakes Water Quality*. By this time the commission was devoting increased attention to the problems of air pollution in its reports, and would continue to do so in the future. The release of the Great Lakes Water Quality Board and Science Advisory Board 1978 annual reports in July 1979 resulted in a sense of urgency and more widespread awareness about acid rain. The conclusions of these reports were discussed at a public meeting that received much media attention. At the meeting IJC officials and board scientists stated that if acid rain was not controlled: more than fifty thousand lakes in the United States and Canada could be devoid of fish and other life forms by 1995; vast areas of prime forest land could be rendered economically unproductive for decades; and drinking water and fish supplies for millions of people could be contaminated by poisonous metals extracted from water pipes or rocks and sediment by acidified waters.[62]

These concerns were included in the commission's landmark *Seventh Annual Report on Great Lakes Water Quality* in 1980. This report led to subsequent discussions in the U.S. government about invoking Section 115 of the Clean Air Act. Four days before he left office, U.S. Environmental Protection Agency Administrator Douglas Costle announced that, based on the IJC report and recent actions taken by the Canadian government, the EPA might be justified in requiring certain U.S. states to cut air pollution contributing to the acid rain problem in Canada.[63]

In 1980 the commission expressed concern in a letter to the governments about a proposed U.S. coal conversion program. The commission noted that the program would have serious transboundary implications and that, without adequate remedial measures, adoption of the program would increase transboundary acid precipitation. The commission also advised the governments to apply the common approaches toward solving environmental problems that had been developed over the years so that timely measures to control and reduce acid rain could be implemented.[64]

Despite these urgings, Canadian and U.S. actions to alleviate the problems of acid rain have not been forthcoming. In 1982 the IJC noted in its first biennial report under the 1978 Great Lakes Water Quality Agreement that "it is time to give them this overdue attention."[65] Prospects for government approval of third-party action outside the context of an acid rain accord presently seem remote because of a variety of political, economic, and scientific uncertainties. The commission has not been included in current bilateral research efforts, and government officials admit that they do not foresee a role for the IJC, or some other binational third party, until after the signing of a North American acid rain accord. The question that arises, then, is to what extent is it possible for the IJC or a similar organization to contribute to the formulation of such an accord?

The IJC and the Resolution of the Acid Rain Issue

If the IJC were to receive a reference concerning acid rain, it would be after the governments had negotiated a treaty.[66] Thus, the role of the IJC would not be to provide information to help further ongoing negotiations, but to monitor and possibly coordinate bilateral efforts to implement an agreement. This is the function performed by the commission under the Great Lakes Water Quality Agreements and, on a smaller scale, under the Michigan-Ontario Memorandum of Understanding. It is possible, therefore, that the governments will ask the commission to perform a similar function once they reach an agreement on acid rain. To date, however, specific uses for the commission have not been discussed in the bilateral negotiations because officials believe that an acid rain accord is still several years away.[67]

The prospects are remote for creation of a new third-party institution based on the IJC model to monitor and report on the implementation of an acid rain agreement. Creating a new organization would take a great deal of time and involve a number of risks. There is no guarantee that the governments could agree on what form the organization would take, much less that they could agree on the extent of its powers. Despite the occasional governmental objections to past commission activities, the practice of having the ijc oversee the implementation of U.S.-Canadian agreements is worthwhile, and should be continued if suitable conditions prevail.

Just what would be the proper conditions for ijc participation in a bilateral acid rain accord? One can only speculate about a possible role if the negotiated agreement should, to some extent, resemble the Great Lakes Water Quality Agreement of 1972. And there are, indeed, some similarities between what the governments were trying to do then and what they are striving for now. In both cases the governments seek to achieve pollution reduction through jointly agreed-upon objectives and schedules of pollution control. Once clear goals and a timetable for achieving them are approved, it would be appropriate and feasible for the ijc to monitor and report on bilateral acid rain abatement activities.

The purpose for building a third party into a bilateral agreement would be to enhance confidence in each country's efforts to control acid rain. Using the ijc for this purpose would be advantageous for a number of reasons, many of which have been written about elsewhere.[68] Most notably, the commission is fully established and has accumulated valuable experience in dealing with transboundary water and air pollution problems. The commission also has the capability to engage in fact-finding and formulating technical and policy-oriented reports. Moreover, through its boards the ijc is accustomed to dealing with government scientists and policymakers. By using the ijc, the governments can avoid the inevitable growing pains that accompany a new, untested organization.

Another advantage to using the ijc as a treaty (or agreement) monitor is that the acid rain issue has already been incorporated into the commission's mission. Since 1978 the ijc has considered the effects of air pollution on aquatic and land-based ecosystems, a primary area of interest for policymakers attempting to understand the problems caused by acid rain. The commission's experience with acid rain-related problems would enhance its ability to monitor and report on governmental efforts to stop acid rain.

Problems

Although the experience of the commission is generally considered its strongest attribute, it may also be a weakness. The ijc has developed long-established ways of doing business and deeply ingrained relationships with governments. Until now the IJC has dealt primarily with transboundary pollution problems on a site-specific or regional basis. Because the acid rain problem may be continental in scope, use of the IJC as a bilateral monitor could force the commission to alter its institutional perspective, and might even create the need for organizational restructuring. Either would create internal upheaval and possibly affect the commission's ability to meet its existing commitments to boundary waters management and ongoing references.

Even if the IJC were able to adjust successfully to a broadened role under an acid rain agreement, could the governments do so? Both Washington and Ottawa are reluctant to see the commission become a "super environmental protection agency." Government actions through the years have been restrictive by denying the IJC enforcement powers, reducing the importance of the regional office, and establishing the BRCG outside the aegis of the commission. These moves made it clear that neither country wished to grant new jurisdictional authority to the IJC. Memories of the commission's perceived environmental activism of the mid-to-late 1970s linger, and governmental sensitivity to IJC activities could severely restrict the terms of an acid rain reference. Moreover, the declining level of contact between the commission and the governments—especially in the United States—could reduce the impact of commission reports.

Although government officials and informed observers expect the commission to receive monitoring and reporting responsibilities as a result of government negotiations, what this actually means is unclear. How would the commission perform a monitoring function? Several options are available.

1. The IJC could fund and operate an air pollution–acid deposition monitoring system of its own. Under this option the IJC would use the data it collected to assess the effectiveness of governmental efforts to reduce industrial and auto emissions, which are the most likely causes of acid rain.
2. The commission could fund and operate a set of "core" monitoring sites which would supplement an overall monitoring system established by the governments.[69] Under this option the IJC would supervise governmental implementation efforts and would operate monitoring facilities as both a check on government implementation and a protection against unforeseen cutbacks in government funding of their own monitoring systems.
3. The commission could conduct no air pollution–acid deposition monitoring itself, but concentrate instead on oversight of govern-

mental monitoring and other agreed-upon acid rain abatement activities.

Complete IJC funding and operation of a monitoring system would be inconsistent with its historical role. Neither government is likely to agree to such an expansion of IJC activities. At present the commission's monitoring activities under the Great Lakes Water Quality Agreements and the Michigan-Ontario Memorandum of Understanding consist only of IJC supervision of municipal and federal activities; and this supervision is carried out primarily by board members on government payrolls. As a general rule, the IJC synthesizes and evaluates information; it does not use its own resources to generate it. In contrast, IJC-funded monitoring would require a large expansion of commission resources and would duplicate existing governmental efforts.

An operational role would also diminish the commission's capability to maintain the outlook of an objective third party. If the IJC were to fund and operate a monitoring system, its traditional role of focusing public attention on and responding to transboundary concerns as an independent observer would be compromised. As an important part of the agreement implementation process, the commission would bear at least partial responsibility for the success or failure of U.S.-Canadian efforts to abate acid rain. The commission would no longer be in a position to report on the adequacy of government efforts to reduce airborne pollutants because the adequacy of its own monitoring system would be open to criticism, especially given the relatively low sophistication of many present monitoring systems and the difficulty of tracking the pathways of airborne pollutants.

Commission operation of a core monitoring system, while supervising the pollution abatement and monitoring efforts of the governments, is also unrealistic. Again, this option would expand commission resources beyond acceptable levels and would duplicate government efforts. It offers no real protection from cutbacks in government monitoring activities because the governments could just as easily cut back the IJC budget as their own—and they might be more inclined to do so.

The most realistic option is to have the commission concentrate on oversight of governmental pollution monitoring (emission and deposition), as well as other agreed-upon acid rain reduction measures. By remaining an observer and analyst, the commission would retain its traditional role. Moreover, it would continue to evaluate government activities and recommend ways to improve them. Through negotiation the governments would set the standards by which the commission would evaluate binational activities aimed at reducing acid rain; in its evaluations the commission would determine if the United States and Canada are meeting their own standards, and if the standards remain adequate over time. All this would be possible with comparatively minor alterations to the IJC organizational structure.

Reporting

In the near term, acid rain promises to remain a politically charged, bilateral issue. One consequence will be continuing governmental sensitivity to IJC (or if the occasion should arise, other third-party) reports on federal compliance with a negotiated agreement. An important issue to be considered, then, is how, when, and to whom commission reports on acid rain would be released. Government negotiators must also decide what issues the commission will be free to address, and whether it will be allowed to address them publicly.

One option would be simply to grant the commission the same freedom it enjoys under the Great Lakes Water Quality Agreements. The IJC would be allowed to report with authority and publicly on: the quality and adequacy of ongoing government air pollution reduction efforts; the results and significance of other government air pollution and acid deposition efforts; when and where violations of, and lack of commitment to, the acid rain agreement are apparent; and needed remedial action.

Whether this would ensure a maximum bilateral effort to reduce acid rain is an open question. Of course, the desired effect of commission reports would be to rivet public attention on identified problems until corrective national action was taken. On the other hand, a publicly issued commission report that highlights implementation shortfalls could either inflame bilateral relations or create government hostility toward the commission.

Unfortunately, other available courses of action, such as restricting the distribution of IJC reports or narrowing the range of commission oversight, would at the same time reduce governmental accountability and reduce commission effectiveness. Consequently, permitting comprehensive public reports to be issued is the most appealing option, especially given the relatively broad public awareness and concern about acid rain. Whether the governments will brave the potential embarrassment that public third-party reports can generate is at present an unanswerable question.

The granting of enforcement powers to the commission is unlikely. There is no reason to expect that the traditional government reluctance to do so will change. Moreover, granting the commission enforcement powers could be counterproductive. Enforcement powers, like operational monitoring responsibilities, would lock the IJC into existing ways of doing things and necessitate a large growth in commission staff. In addition, granting the right of enforcement would make it possible for the commission to become a scapegoat in the event efforts to reduce acid rain fail.

Given the exceptional delicacy of the acid rain issue, government negotiators should consider insulating the commission from political pressure. For example, consideration should be given to establishing fixed terms for U.S. commissioners. (This is already done in Canada.) Multiyear budget appropriations would also be an easy method to ensure IJC independence.

On the other hand, the commission will have to be constantly alert to the sensitivity with which the governments will receive its reports. It will have to stay within the range of activities authorized by the agreement and avoid initiatives that would resurrect government hostilities and charges of environmental activism. Board activities will have to be closely controlled, and inflammatory statements at public hearings should be avoided.

Finally, and most important, the commission should attempt to elicit governmental response to its findings by emphasizing short- and medium-term specific recommendations. To increase the impact of its long-term recommendations the IJC should attempt to reestablish regular high-level contact with senior policy officials at the Departments of State and External Affairs. In the final analysis it will be the governments that determine the success or failure of commission endeavors. If the commission is called upon to monitor binational acid rain cleanup efforts, maintenance of good working relationships with Washington and Ottawa must be a top IJC priority. Given the extent of current bilateral disagreement and confusion about acid rain, it promises to be a difficult task.

Chapter Nine
Supporting Structures for Resolving
Environmental Disputes among
Friendly Neighbors

Introduction

The past history of U.S.-Canadian environmental relations demon-
strates the utility of alternatives to, or supporting structures for, formal actions by the federal governments. Actions by courts, regulatory
intervention, bureaucratic relations at both the federal and subfederal
levels, and informal organizations such as the Canada–United States
Interparliamentary Group all provide means by which each country
can spur development of favorable policies in the other.

This chapter examines previous uses of each of these structures in
connection with the environment. These earlier experiences lead to
the following conclusions:

Informal forums for policymakers on both sides of the border can
contribute to the solution of transboundary pollution issues by
increasing their knowledge and understanding of each other's
positions.

The use of joint committees and commissions such as the International Joint Commission have facilitated resolution of disputes
involving other types of pollution.[1]

Existing cooperation between subfederal jurisdictions could be
expanded to increase coordination of policies that affect transboundary environmental concerns.

Formal channels ranging from court suits to intervention in regula-

This chapter was written by Andrew Morriss.

tory proceedings are available to increase the attention given foreign concerns by both decision makers and the public.

A number of steps, some largely symbolic but still important, exist for both governments to demonstrate their willingness to bargain over acid rain, ranging from equal-access reforms to changes in substantive requirements in statutes.

Our analysis suggests that use of supporting structures provides a means for both increasing the chances of a federal-level formal agreement and affecting immediate policy decisions that relate to acid rain.

Although Canadian-U.S. environmental disputes have been labeled "environmental" only since the late 1960s, they existed long before that. A "long standing tradition of 'regional action' with the United States to curb transboundary pollution" developed in accord with the general 'good neighbor' relations of the two countries.[2] The 1909 Boundary Waters Treaty set the general tone of these relations, concentrating on water pollution.[3] Those disputes which did involve air pollution were limited to short-range transport of pollutants.[4] Examination of that history of relations provides useful lessons for the acid rain dispute.

Informal forums, lawsuits, and regulatory intervention are only a few of the channels outside "normal" federal-federal relations available to a government wishing to influence the domestic policies of a neighbor. These channels can influence policy directly, as intervention does, or indirectly, as a lawsuit does. They are not a substitute for direct federal-level agreement, but they do offer possibilities for nudging the domestic policy debate in the direction desired. In the case of good neighbors like the United States and Canada, these structures are more developed than between less friendly neighbors. The possible use of these well-developed supporting structures is of great interest for the acid rain debate.

All of these structures are available in both the United States and Canada. It is only in the United States, however, that these avenues are likely to succeed with any frequency. A host of problems awaits the potential participant in the Canadian legal and administrative systems. Thus Canada has a greater opportunity to affect U.S. policy than the United States has to affect Canadian policy.

Because the acid rain conflict to a large extent is more a conflict between regions than between nations, it is important that the U.S. victims have some access to the Canadian policy process to influence results in those cases where the United States is downwind. Canada as a whole may take more notice of the United States as a whole than vice versa, but the relationship may be more equal between Montana and Saskatchewan, or New York and Ontario. At this state-provincial level many crucial decisions are made relating to acid rain and to environmental protection in general. Consequently, political divisions on both sides of the border need access to their counterparts. It is here that the lack of ability to influence substantive policy-making in Canada is most critically absent.

The role of this chapter is to examine those structures that seem suitable or that have been frequently mentioned in connection with acid rain. The chapter is divided into three sections: formal remedies, informal means of pressure, and subfederal relations. The first section outlines mechanisms involving formal participation in the neighbor's policy-making process. The second examines structures, such as the Canada–United States Interparliamentary Group, that provide opportunities for informal discussion and consultation among policymakers. The last investigates nonfederal avenues of communication.

Formal Remedies

There are four formal mechanisms for influencing policy outside of diplomatic channels: use of the courts, intervention in regulatory proceedings, use of the international provisions of the two national Clean Air Acts, and equal access to courts and administrations of the other country.

Use of the Courts

The court system is oriented primarily toward compensating wrongs, but it is also a decision-making process that must relate its decisions to both public policy, through the precedential value of its opinions, and to the individual facts of the cases before it. Court-made law has been and will continue to be a major force in the development of environmental law.[5]

Private-law solutions to transboundary problems are initially attractive because of the deficiencies and failures of public international law.[6] They serve a valuable function in and of themselves, providing compensation in cases where one clearly identified party has harmed another. Although this is difficult to prove in most environmental law cases, and particularly so with acid rain, improvements in tracing techniques could make it much easier in the future. More important for the development of policy, private damage actions provide an opportunity to spur the legislative and executive branches of government to action to remedy a situation, as was the case with water pollution and the 1899 Refuse Act in the United States (30 Stat. 1151).

Successful tort cases build pressure for a higher-level remedy for the problem. Large damage awards against polluters create incentives for government action to alleviate the burden placed on the individual polluters found liable, through greater sharing of abatement costs and reducing polluters' incentive to delay resolution of the problem. Even unsuccessful lawsuits provide publicity and an opportunity to encourage debate over a policy issue. Unsuccessful efforts may even be more effective in garnering publicity than successful ones, because being defeated on what may appear to be a technicality, and leaving the court empty-handed, may generate more sympathetic coverage than a multimillion-dollar victory.

Nations are often unwilling to push claims, either for political and diplomatic reasons or because they are reluctant to assert a principle of international law that could be turned against them at a later date.[7] In such cases private damage actions provide an opportunity for affected individuals and groups to attempt to remedy their situation and/or affect policy outside their own federal government. The suits are also useful where the state or nation is unsuccessful in pushing its citizens' claims.

Problems with Lawsuits   Many barriers exist to cross-jurisdictional lawsuits, and attempts to grant citizens equal access to the courts of another country are aimed at removing some of them. It is important to remember that reforms of this kind, even if perfectly designed, will do no more than provide rights in another jurisdiction commensurate with the rights enjoyed by that jurisdiction's own citizens. However, having equal access to Canadian courts provides few environmental rights. Moreover, "attitudes of self-interest and national autonomy regarding environmental problems are shared by judges as well as by legislators and bureaucrats."[8]

Because there have been relatively few international lawsuits between U.S. citizens and Canadians, interstate and interprovince lawsuits have developed the jurisdictional and standing rules that are the first hurdles in any private action.[9] Because Anglo-American practice begins with the courts, the first problem is jurisdiction of the court (rather than choice of law, as it is in continental systems).[10] In the United States there is little problem securing jurisdiction over foreign defendants, although enforcement of a judgment in a foreign country could be a problem. The two tests for jurisdiction in the United States are due process and forum statutes. The former is essentially a requirement that the defendant have some connection with the state whose court is trying him.[11] The latter is met by "long-arm statutes" in most states, which endow their courts with jurisdiction over non-resident defendants.[12]

Jurisdiction does present a problem in Canada, even with pollution suits where the polluter is in one province and the damage occurs outside the province. Canada follows an interpretation of a House of Lords decision, *British South Africa Co.* v. *Companhia de Moçambique*, which virtually prohibits public interest lawsuits.[13]

There are three possible interpretations of the *Moçambique* rule: (1) domestic courts have no jurisdiction to adjudicate title or right of possession of foreign land; (2) domestic courts have no jurisdiction to adjudicate upon title or grant damages for trespass to foreign land; and (3) domestic courts have no jurisdiction to entertain any actions involving damage to foreign land, whether or not title is at issue.[14] Canadian courts have chosen the last, and most restrictive interpretation. This is undoubtedly one of the reasons why Canada has never had an interprovincial nuisance action.[15] By contrast, the rule in the United States is similar to the least-restrictive first interpretation.[16]

Another major problem for a lawsuit within Canada is standing. In cases of public wrongs, the classification of most environmental cases, only the attorney general or someone acting on his behalf has standing to sue in the public interest.[17]

Another problem for the potential litigant in Canada bringing a transboundary environmental pollution case is Canada's archaic class-action laws. All but Quebec's are over one hundred years old.[18] The major deficiencies in these rules are: the stipulation that the plaintiffs must have virtually identical interests, not just similar interests; the lack of safeguards to protect absentee class members; the prohibition on damage actions where individualized assessments are needed; and the rules governing court costs and legal fees that make class actions expensive and risky for plaintiffs.[19]

When plaintiffs have sued to remedy a private wrong, most pollution actions have been held as not meeting the requirement for identical interests. In one case fishermen were denied class-action status in a suit over ocean pollution. The court held that the fishermen's right to the fish was no greater than the general public's, and consequently denied them status.[20] Similarly, the Supreme Court of Ontario dissolved an injunction to protect fishermen and cabin owners from pollution of the Spanish River by a pulp factory.[21]

Other problems exist for lawsuits in either jurisdiction. Courts are reluctant to grant relief other than money damages in transboundary cases.[22] Even money judgments may be difficult to enforce because enforcement of foreign judgments is determined by reciprocity; the judgment will be enforced only if the judgment court obtained jurisdiction in a manner that would be acceptable in the other country.[23] Identifying the responsible parties is difficult, particularly in long-range transport cases. It can be difficult to find assets in the jurisdiction to satisfy a judgment when there is no enforcement elsewhere. The cost of large lawsuits, particularly multiparty ones, is prohibitive in many cases. It may also be impossible to join all the defendants in a single lawsuit, thus requiring several suits, with resulting additional expense and effort.[24]

These points make environmental suits in Canadian courts by foreign plaintiffs almost impossible. The *Moçambique* rule, the fact that costs can be assessed to the loser, and causation-and-proof problems all lead to the conclusion that "the likelihood of success under current Canadian law is slightly more than zero."[25]

Michie v. Great Lakes Steel   This is the single reported case of Canadian residents suing in U.S. courts for damages from transboundary air pollution.[26] In *Michie v. Great Lakes Steel Division, National Steel Corporation*, the Sixth Circuit upheld a federal district judge's denial of the defendant's motion to dismiss a suit by thirty-seven residents of LaSalle, Ontario, against three U.S. corporations for damage from transboundary air pollution.[27] The suit was an initiative of Windsor District Pollution Probe, a Canadian environmental group.[28] The case lasted five years and resulted in a settlement of $105,000 and an agreement by the defendant companies to spend $4 million on pollution control equipment. The original suit had requested $3.7 million, $3 million of which was punitive damages.[29]

The procedural history of the case is brief. One temporary appeal was made by the defendants on the court's interpretation of Michigan nuisance law. The trial court was upheld by the Sixth Circuit, and the Supreme Court declined to hear the case. Perhaps the most interesting aspect of the case is that little notice was taken of the fact that the plaintiffs were Canadians. Thus, "the *Michie* decision clearly reaffirms the willingness of American courts to preside over the claims of transboundary plaintiffs who have suffered damage to realty. Such a willingness is of considerable importance in view of the reasonable assumption that most cases of cross-boundary damage will fall within this category."[30]

Utility of Court Actions   Where not barred by procedural barriers or substantive problems, lawsuits offer the potential for the traditional remedies both of injunction and of damages for the individuals injured. More important for policy purposes, they offer a chance to influence policy through the judicial branch and in a manner that generates publicity for the cause of the plaintiff. Perhaps most important, the avenue they provide is independent of federal action on either side.

Regulatory Intervention

Public participation in agency decision making is another way of formally influencing the policy process. Intervention can occur in both state and federal regulatory proceedings. Canada and the United States present different obstacles to potential intervenors.

U.S. Intervention in Canada   The right of access to government information is extremely limited in Canada but relatively free in the United States. For example, the Canadian government has denied the Canadian Environmental Law Association access to the results of pesticide residue tests and a list of insecticides containing vinyl chloride.[31] Canada has no equivalent of either the U.S. Freedom of Information Act or the Administrative Procedure Act. Canada's environmental impact statement (EIS) process is also weaker than the U.S. counterpart because public participation in the EIS process is discretionary.[32] Thus, the ability of nongovernmental organizations or private citizens to intervene effectively is weakened.[33] Environment Canada (the federal department in charge of environmental policy) has said it would provide access to: (1) published departmental scientific papers in departmental publications; (2) routinely collected data on ambient levels of pollutants within six months of collection; (3) data collected pursuant to federal regulations; and (4) information collected jointly with the provinces or international bodies if the other parties agree to its release. A general caveat provided that Environment Canada would determine the form of the data to be released—either in the form collected or the form published. In spite of difficulties associated with access to information, however, Canadian governments are willing to allow standing to intervene before public boards. John Swaigen cites only one reported case of an intervenor being denied standing by a board, and that decision was overturned by the courts.[34]

Litigation after the regulatory decision has been made is much more difficult in the Canadian system than in the United States. This is illustrated by a comment the Ontario deputy minister of the environment made before a U.S. Environmental Protection Agency (EPA) hearing:

"[A regulation in Ontario is] subordinate legislation having the same effect as a statute. It is not subject to any statutory appeal. If it is within the statutory power and made in good faith, it is almost unassailable in the courts."[35] Canadian environmental groups are therefore more oriented toward public education than court action, unlike their U.S. counterparts.[36] Thus, U.S. environmental groups would probably be granted standing before administrative tribunals in Canada,[37] but they would be unlikely to contest any results successfully.

Canadian Intervention in the United States   The situation is quite different in the United States, where Canadian groups quite commonly intervene. Canadian environmentalists have appeared before the Nuclear Regulatory Commission to argue against nuclear power plants near the border,[38] before a Washington State court to argue against a tanker route from Alaska,[39] and before a federal court to argue against the Alaska pipeline,[40] as well as before numerous boards and other forums in connection with water pollution and usage issues.

The provincial government of Ontario has also become quite active in intervention in recent years, primarily over acid rain-related issues. Ontario filed a petition with the U.S. EPA requesting that the agency deny a proposed relaxation of emissions standards for seventeen power plants in six states.[41] Ontario sought to broaden New York's and Pennsylvania's actions to deny the proposed relaxation under the interstate pollution provisions of the Clean Air Act to include international pollution. The EPA denied Ontario's request but allowed the province to make a presentation during the two days of hearings on issues that would assist EPA in making a decision on the two states' petition.[42] Ontario's deputy minister of the environment testified along with several provincial experts and the province's local legal counsel.

The deputy minister concentrated his testimony on describing Ontario's policy approach to acid rain and urging adoption of a similar policy by the EPA.[43] In its original petition Ontario cited a wide variety of international documents in support of its position, including the Memorandum of Intent on Transboundary Air Pollution, the Great Lakes Water Quality Agreements, the Trail smelter agreement, Principle 21 of the 1972 Stockholm Declaration on the Human Environment, and the United Nations Convention on Long Range Transport of Air Pollution.[44]

Ontario has also intervened on the state level with a presentation to the Michigan Air Pollution Control Commission. In its petition Ontario stated that it was provincial policy to accord U.S. citizens and agencies the same rights of intervention as Canadians in regulatory proceedings involving transboundary pollution interests.[45]

The 1974 *Wilderness Society* v. *Morton* decision established the right of foreign groups to challenge in court U.S. regulatory decisions that affected them. There have been very few decisions since then, however, that have addressed the rights of foreign nationals in the U.S. environmental policy decision-making process. In *Wilderness Society* Canadian environmental groups intervened in opposition to the route selected for the trans-Alaska pipeline. Specifically, both the Canadian and U.S. intervenors sought to raise the issue of an alternate route through Canada, alleging their proposed route was less environmentally damaging, particularly to Canada, than the route selected. In one of the many appeals that arose from this case, the court noted that the original appeal of the agency decision by both sets of intervenors "helped focus attention in Congress on the major issue raised—the relative merits of a trans-Canada versus a trans-Alaska route."[46] The Canadian intervenors were accorded full equality with the U.S. intervenors, including the award of attorneys' fees.

People of Enewetak v. *Laird*[47] interpreted *Wilderness Society* to hold that foreign nationals have the right under the National Environmental Policy Act (NEPA) to an EIS on the impacts "when their environment is endangered by federal actions."[48] The Council on Environmental Quality and the State Department disputed for many years the applicability of NEPA to federal actions affecting foreign environments. None of the relevant court cases resolved the issue, however, as they either never reached the issue or approved a settlement including an EIS.[49] Canada has been singled out by the courts as having a "special relationship" with the United States which requires that federal actions with environmental repercussions outside the United States be subject to environmental impact statements.[50] An executive order by President Carter mandated consideration of foreign environmental consequences by several different methods.[51]

In the most recent court action surrounding Canadian intervenors, *Swinomish Tribal Community* v. *Federal Energy Regulatory Commission*, Canadian and U.S. intervenors opposed the increase in size of a dam that would flood part of British Columbia.[52] Both sets of intervenors participated fully in the hearings on the raising of the dam, with more than a dozen witnesses testifying on the Canadian impact.[53] The International Joint Commission (IJC) had earlier been called in to assess the impact of the dam on Canada, and its recommendations were made part of the final EIS. The IJC had imposed the condition that the province be compensated for its flooded land, and an agreement was reached to do so. The Canadian intervenors, although independent of the U.S. intervenors, shared an attorney with them and argued that an independent EIS on the Canadian impact was required, because mere incorporation of the IJC report was insufficient. The appeals court rejected this argument.

Benefits of Intervention   Intervention provides an opportunity for foreign nationals and subfederal units to participate directly and formally in the policy-making process.[54] This participation yields two sets of benefits to the intervenor. First, it has direct impact on the foreign policy process, an opportunity to present formally information and views that must, in the United States at least, be considered by the agency in making its decision. Second, publicity can be generated from an intervention to present an intervenor's point of view to the foreign public, thereby indirectly influencing policy formation. If, as the Canadian Senate subcommittee concluded, a major part of the problem in the United States is the "appalling ignorance and lack of concern" over acid rain, then intervention offers an easy way to reduce that ignorance.[55]

Intervention need not be limited to environmental decisions. Agencies such as the National Energy Board of Canada and the U.S. Federal Power Commission make decisions that affect acid rain as well. Intervention there opens the door to educational efforts on the link between energy policy and acid rain. The flip side of the intervention coin is that a potential backlash could be generated—from the foreign public over perceived interference in their government's activities, from the agency and staff involved, and from the parties against whom the intervention is made. Ohio, for example, might be less willing to cooperate in the future if Ontario repeatedly intervened against it in regulatory proceedings.

The Clean Air Acts

Although the possibility for intervention in regulatory proceedings exists in both nations, most interventions are of limited benefit to the intervenor because their scope is usually limited to domestic impacts of the action being considered, except where a direct foreign environmental impact can be shown, as was the case in *Wilderness Society*. In the case of Ontario's proposed revision of the Ohio State Implementation Plan (SIP), Ontario was limited to presenting information that would assist EPA in assessing the environmental impact of the relaxation on the United States. In order to have maximum effect on policy, the impacts on the foreign jurisdiction must be considered. Moreover, once the proceeding is open to include foreign impacts, the door is open for lower levels of government such as municipalities and other concerned groups to present information. To accomplish this, however, provisions of the two federal Clean Air Acts must be brought into play. Regardless of their other differences in structure and style, both the U.S. and the Canadian acts provide for inclusion of foreign environmental impact in the policy process under certain similar circumstances.

U.S. Clean Air Act Provisions   The international provisions of the U.S. Clean Air Act are of recent origin. The original Clean Air Act passed in 1955 made no mention of either interstate or international air pollution. The problem of interstate pollution was first addressed in the 1963 amendments, which authorized the Department of Health, Education, and Welfare to call a conference of affected air pollution agencies. The secretary was given minimal powers to produce an agreement, but no power to set standards. In 1965 a similar provision was added for international pollution. The 1977 amendments provided for the modification of the SIP, rather than the conference, as an implementation device.[56]

In the 1977 amendments Section 115, International Air Pollution, was added to the act. It requires the states to take export of pollution to foreign areas into account when formulating their SIPs. Or, if states fail to do so, the EPA can order a new SIP that takes transboundary pollution into account. In addition, the act requires EPA to give notice to affected parties, including foreign nations, of changes in emissions limits.[57]

Two actions are required to trigger Section 115 of the U.S. Clean Air Act: (1) the administrator must have reason to believe that air pollution from the United States "may reasonably be anticipated to endanger public health or welfare in a foreign country"; and (2) the administrator must find that "the United States [is given] essentially the same rights with respect to the prevention or control of air pollution occurring in that country as is given that country by this section."[58] Figure 9-1 lists the various conditions under which a finding of endangerment may be made. Once the section is invoked, the administrator notifies the governor of the state causing the emissions to revise the SIP to eliminate the threat. The affected foreign country is invited to appear at the SIP revision hearing.

Several areas of uncertainty exist with regard to these procedures.

Figure 9-1.   Finding of Endangerment under
Section 115 of the U.S. Clean Air Act

United States		Canada	Finding of endangerment
violation	⟶	violation	automatic
no violation	⟶	violation	discretionary with United States
violation	⟵	violation	automatic
violation	⟵	no violation	discretionary with Canada

Note:
⟶ : direction of emission
violation: violation of laws or standards
endangerment: finding under Section 115 that health and welfare are endangered.
Source: Environmental Mediation International, *Use of Section 115*, p. 43, chart no. 1.

First, it is unclear just who may petition the administrator to begin the process. The statute uses the phrase "any duly constituted international agency" but does not define the term.[59] It is unclear whether this includes the affected country's air agency as opposed to just the federal government, and if it includes truly international bodies like the IJC[60] or the Organization for Economic Cooperation and Development. It does suggest that private citizens and organizations would be denied access to the system.

Second, it is uncertain just who would be invited to participate in the SIP revision—the question of citizen and private organizational participation is left undefined.[61]

The concept of reciprocity is also unclear. Viewed narrowly, it could be defined as a requirement that provisions must exist in the foreign jurisdiction identical to those in the Clean Air Act. A broad view would require only that the same substantive result be achieved through whatever mechanism the other jurisdiction provided.[62] Considering the differences in distribution of powers among the federal and subfederal systems of the two countries, it is important for Canada that the final definition be closer to the latter extreme.[63] The most likely definition is one of rough parity, where the Canadian federal government has the authority to ensure performance by the relevant actor,[64] without restriction on the source of the authority to the Canadian Clean Air Act.[65] Under the Carter administration, EPA extended the reciprocity concept to subfederal units of government that possess independent power to control air pollution.

Use of Section 115

In December 1980 the Canadian Parliament passed Bill C-51, which amended the Canadian Clean Air Act to provide for reciprocity with the United States[66] and gave Ottawa the power to regulate in cases of transboundary pollution. In an extremely unusual move, C-51 was debated and passed, unanimously, in less than one day.[67]

The EPA has not always complied with Section 115. It twice approved relaxations in SO_2 emissions limits without informing the affected states or Canada.[68] But following the action by the Canadian Parliament, Administrator Costle concluded that the reciprocity requirement had been met. He stated that the 1980 annual report of the IJC met the first requirement for Section 115 action, and that the bill's passage by Canada met the second requirement. This was announced in a January press release and again in a letter to Senator George Mitchell of Maine. In the letter Costle said: "In my view the amendments to the Canadian Clean Air Act do give adequate authority to the government of Canada to provide essentially the same rights to the United States as Section 115 provides to Canada. . . . Both statutes allow the state or province, as appropriate, to take actions to remedy air pollution affecting a foreign country. If the state or provincial government fails to develop an adequate remedy, the federal government is authorized to establish emissions limitations."[69] Ohio and two Ohio utilities filed petitions in the federal court of appeals asking the court to review Costle's action and set it aside.[70] It is not clear just what effect, if any, Costle's remarks had.

The Reagan administration has interpreted Costle's statements to mean that research is required to determine if a problem exists, a position consistent with the administration's general views on acid rain.[71] Because this research is already under way under the auspices of the Memorandum of Intent (MOI) work groups, no additional steps are planned. EPA has indicated that to actually trigger Section 115, "proof" would be required of a definite link between specific sources and acid precipitation by either the MOI groups or by a body such as the IJC.

To date, intervention in Clean Air Act proceedings has not been a promising way to address the problem of acid rain. The present method of controlling air pollution under the U.S. Clean Air Act is through the setting of ambient standards. The standards are ineffective in dealing with long-range pollutants like acid rain because they focus on ground-level concentrations.[72] The standards do not set a goal of reduction of total atmospheric loading of pollutants, but focus instead on achieving local health and environmental goals. The act provides no incentives for states to control in-state sources of pollution in response to out-of-state impacts. Federal authority to force such consideration has proved to be limited.[73] Thus, establishing a means to provide such incentives should be a major goal of any long-term solution.

The revisions of SIPs in several midwestern states demonstrate these problems. New York, Pennsylvania, and Ontario, as well as several other states, challenged EPA's decision to revise and relax emissions limits in Ohio and other midwestern states. After initially failing to expand the hearings under Section 115 to cover the impact on Ontario of the increased pollutants, Ontario presented information to EPA on acid rain generally, and on Ontario's policy approach to the problem. Ontario appeared at the June 1981 U.S. EPA Section 126 hearings in support of the states of New York, Pennsylvania, and Maine in their petition concerning interstate pollution. A complicated set of proceedings arose, with Pennsylvania and New York maintaining that SO_2 and particulate sources in the Midwest were preventing the attainment of national ambient air quality standards.

In March 1984 New York, Maine, Vermont, Rhode Island, Connecticut, and Massachusetts, together with several environmental groups and U.S. citizens, filed suit against EPA, claiming that:

1. the Administrator has violated his mandatory duty under Section 126 to issue a final decision on the petitions regarding interstate air pollution by the deadline (within sixty days) specified in the statute; and
2. the Administrator has violated his mandatory duty to determine which states are contributing to air pollution which endangers the public health and welfare of Canada and to give notice to the governors of such States to revise the State Implementation Plans in order to prevent or eliminate harm.

In December 1984 EPA denied the Section 126 petition and the decision was appealed to the U.S. Circuit Court of Appeals. As a result, U.S. Federal Judge Norma Holloway Johnson ordered the EPA to set in motion the necessary processes to reduce acid rain emissions. During the appeal process Ontario obtained party status and filed a brief. The U.S. district court reserved judgment, and in September 1986 the U.S. Court of Appeals for the District of Columbia Circuit rendered its decision to reverse and remand the order to district court with instructions to dismiss. This decision was subsequently upheld by the U.S. Supreme Court.[74]

Canada   In Canada the federal role in air pollution control "is one of guidance and demonstration to the more autonomous provinces."[75] The provincial programs, the mechanisms where actual pollution control occurs, tend to be flexible rather than prescribing requirements, and the emphasis is on industry-government cooperation.[76] This is in sharp contrast to U.S. practice, where the primary mode is adversarial. Provincial enforcement is often a low priority and consequently suffers from small staffs and smaller budgets. Even successful prosecutions of Canadian environmental laws do not provide much of a deterrent. A survey of cases under Ontario's environmental acts, for example, disclosed that the average fine was $1,000 (Cn) and that the highest to date on a first offense was $4,000 (Cn), despite the fact that the maximum fines were $5,000 (Cn) for the first offense and $10,000 (Cn) for subsequent offenses.[77] Thus one commentator concluded that much of the growth in Canadian environmental law has been paper growth.[78]

Throughout the 1970s the provinces have steadily broadened the scope of their activities in the environmental field.[79] The provinces continue to hold a wide variety of views on environmental issues, attributable to their varying positions in energy development. The result of the provincial focus of environmental controls has been fragmentation of responsibility.

The 1980 action by Canada in response to Section 115 of the U.S. Clean Air Act enlarged the powers of the federal government, which previously had been authorized only to set emissions standards necessary for compliance with international air pollution agreements.[80] However, the extent of the Canadian federal government's powers to actually implement international agreements in areas where provinces have jurisdiction, such as air pollution control, is limited. Implementation in such a case is generally conceded to remain in the hands of the provinces.[81] (For details, see chapter 6.)

Like the U.S. Clean Air Act, the Canadian act is rooted in a 1960s approach to ambient pollution. Unlike the United States, however, public awareness of acid rain is at a high level and the government has taken significant action, but not under the Clean Air Act.

Equal-Access Reforms

Equal-access reforms are steps to ensure that citizens in one country have the same access to legal remedies in a second country as the second country's own citizens. They provide relief from primarily procedural difficulties arising out of multinational litigation. The American Bar Association (ABA) and the Canadian Bar Association (CBA) jointly developed a treaty proposal that provides equal access to each nation's courts for the other's citizens.[82] The treaty developed out of meetings initiated when the CBA officers were approached by the ABA officers in 1975 and 1976.[83] In 1979 the two organizations recommended to their governments treaties providing equal access.

The treaties state that each country's citizens are to be given rights in the other country equal to those of the jurisdiction's own nationals to take part in all existing administrative and judicial procedures to prevent domestic pollution. They also require that environmental organizations must be included along with other domestic groups and that the polluter must provide notice to affected parties of their rights.

The equal-access treaties require no complex determinations of a law or fact, since they simply provide an opportunity for citizens of nation A to present their case to the agencies and courts of nation B as if they were B's citizens. As such, the treaty would be no substitute for direct action on the problem, but it would represent an opportunity for each nation to address concerns in the other. Although the treaty proposal is largely symbolic, given the extent of reciprocal rights now existing, it provides a step toward a more substantial solution and, if adopted, would ensure that the access that is now informally acknowledged in some cases would be formalized. The treaty proposal is not under serious consideration at the moment.[84]

Another avenue for equal-access reform is adoption of uniform legislation by states and provinces. The National Conference of Commissioners on Uniform State Laws, an organization affiliated with the ABA, has approved a draft uniform law, although the full ABA has not yet done so.[85] The CBA and the Canadian uniform-law body have both approved the draft, and thus action there seems more likely in the near future. Several New England states are reported to be considering similar legislation even before the draft is approved by the ABA.

Following the ABA-CBA treaty proposal, the uniform-law conferences of the United States and Canada (private organizations affiliated with the bar associations) met and worked out draft legislation covering uniform access for transboundary pollution cases. In essence, the proposed legislation simply presents the treaty's provisions in a different form. The only substantive differences between the two proposals would be the inclusion of federal-level access in the treaty and fewer actions needed for implementation.

The initiation of Section 115 or Bill C-51 proceedings is not a substitute for equal access. Both provisions address only proceedings under the Clean Air Acts; they do not provide for private actions or intervention in the many other regulatory actions that affect acid rain—such as utilities regulation. Thus even if Section 115 and Bill C-51 are used, which has never yet occurred, provision of equal access is still important.

Summary

These four formal means offer the United States and Canada some opportunities for influencing policy in the other nation. However, the methods offer unequal opportunities to the two nations. For various reasons it is much harder to use each mechanism in Canada than in the United States. Although this may not be critical in the case of acid rain, because U.S. domestic policy is likely to determine that outcome, it is important and has been important in other environmental transborder disputes—Trail smelter, Atikokan, and Poplar River, for example.

In addition, all of these approaches suffer from a common flaw—they are formal remedies and thus take a long time to implement. This could be a problem in the case of acid rain if environmentalist fears of the magnitude of the consequences are true. An equal-access treaty, although it would not address acid rain directly, would be an important step toward resolving the conflict. As such, it would surely be fought by the same interests who oppose an acid rain treaty. Intervention in regulatory procedures and the Clean Air Acts' provisions are subject to similar pressures and take more time than informal consultation between agencies. They are also reactive mechanisms that do not permit policy discussion at an early stage in the development process. Moreover, the formal nature of even "informal" regulatory proceedings would lessen the chances of policymakers to reach an understanding of the effects of their decisions on each other's jurisdiction. Court suits take years to conclude, and often require an extensive commitment of resources by the plaintiff. Thus, although each procedure offers opportunities for solving the acid rain threat, none is sufficient alone.

Informal Influence and Pressure

In addition to the formal remedies discussed so far, various informal means of influencing policy exist at the federal level. Parliamentary exchanges and the normal contacts between neighboring nations provide many such opportunities. With Ottawa and Washington just a phone call or short plane trip away, routine interactions between members of legislatures and executive branches provide opportunities to share information and discuss policies. In this regard the Trail smelter case is frequently cited and commented upon as an example of multilevel cooperation in U.S.-Canadian environmental relations. In that instance both federal governments used many channels to resolve a transboundary air pollution problem, and it thus offers guidance for acid rain disputes. The particular solution to the Trail smelter case, however, is not appropriate to the current dispute.

Parliamentary Exchanges

The Canada—United States Interparliamentary Group (IPG), formed in 1959 to promote informal discussions between U.S. and Canadian legislators, provides a forum where concerns can be expressed.[86] The group has held meetings approximately once per year since it was formed. It is structured informally, with no staff and no minutes of meetings, and it was originally divided into two committees, on defense and economic matters. Environmental matters are discussed frequently by the group,[87] and indeed it was the disputes over the trans-Canada pipeline in 1956 and the development of the Columbia River that helped to create the group in the first place.[88] More recently the IPG has expanded its discussions of environmental matters to reflect the increased interest of legislators in both countries.[89]

The IPG does not resolve issues, but instead shares information and perspectives and thereby reduces conflicts. Because the IPG is so small, and because neither set of participants perceives it to be particularly influential, the group is not a major actor. To encourage the types of exchanges that occur through the group on a wider range of issues, the Parliamentary Centre for Foreign Affairs and Foreign Trade, with support from Canadian foundations, has sponsored conferences between members of Canadian and U.S. legislative bodies.[90] Of the first eight meetings in 1971—72, two dealt directly with environmental matters and two did so indirectly.[91]

In the 1970s, as environmental concerns grew, so did the IPG's involvement with them. By mid-decade transboundary environmental matters had their own committee, but it concentrated almost exclusively on water-related issues.[92] Energy and environmental concerns were combined into a single committee in 1978, when air pollution concerns began to surface in committee discussions. By the next year the U.S. delegation report noted that "there appeared to be a consensus within the committee that our countries should initiate negotiations to establish uniform standards with regard to air quality and environmental protection technology."[93]

Acid rain first appeared in the U.S. delegation's report in 1979, but the discussion was limited to arguments over the direction and magnitude of the problem. The same condition existed in 1980. By 1981, however, the delegates agreed to seek the establishment of an ongoing monitoring group within the IPG to meet twice a year and assess information on acid rain and possible responses to the problem. A U.S. delegate proposed the creation of a North American Air Quality Commission, similar to the IJC, to coordinate enforcement and research efforts, and it was agreed that the matter would be discussed further at the next meeting. Nothing seems to have come of either proposal. In 1982 acid rain dominated the entire conference. Copies of *Still Waters* were distributed to the U.S. members and a film was screened for the full group by the Canadians.

To the extent that the members of the two delegations to the IPG meetings see themselves as representatives of their respective federal jurisdictions as well as national actors, the IPG provides a forum for the discussion of bilateral issues. The low-key atmosphere and lack of structure is well suited to discussion of policy developments on both sides of the border. The relaxed and informal IPG meetings would be an excellent opportunity for a member from Ontario to discuss the problem of acid rain with congressmen from the Ohio valley. This potential was recognized by the Canadian Senate's Subcommittee on Acid Rain, which recommended use of the IPG and other interparliamentary associations and contacts for discussion of the acid rain issue.[94]

At the IPG's sixteenth meeting several measures were suggested, such as having each country's citizens testify regularly before legislative committees on matters of mutual concern[95] and creating a single technically capable official group in each nation to "coordinate national policy toward its neighbor and advise the highest levels of each respective government." These reforms would help reduce the confusion resulting from "the multiplicity of governmental and private institutions in each country" concerned with transboundary issues.[96] Generating such ideas is a valuable function of the IPG.

Agency Contacts

The United States and Canada have extensive relations through mechanisms other than formal diplomatic contacts. Contacts between policymakers through groups like the IPG provide one means for influencing policy in the other nation. Informal contacts also exist between the bureaucracies and executive policymakers on both the federal and state/provincial levels.

Access to the federal level can be channeled through federal contacts on the other side working with their state and provincial bureaucracies. The contact may also be direct, as with the case of Ontario's formal intervention in U.S. regulatory proceedings and retention of its own congressional lobbyist and legal counsel in six states.[97] Provincial activity in this area is extensive, with hundreds of informal contacts between bureaucracies and up through the highest levels on matters from auto licenses to pollution control. Although many of these arrangements are not related to acid rain, as a whole they provide a framework for subfederal units to influence policy in their counterparts through discussion and information dissemination.

In both nations subfederal units of government and other actors have a great deal of influence over the negotiation of formal agreements, particularly when the implementation of the agreement will be a matter for local authorities, as it would be to a large extent with pollution control. With respect to current bilateral negotiations, the subfederal actors often possess leverage over their federal units and can influence the outcome of the negotiations. This is particularly true where the governments are involved in other disputes, as in Canada, where the provinces—especially in the West—are currently engaged in a dispute over the allocation of authority with the federal government, and thus are in an excellent position to exact concessions.

Finally, federal legislators from states or provinces affected by acid rain who have committee assignments with jurisdiction over foreign affairs are in a position to prod along—or slow down—negotiations, particularly in the United States, where Congress is more independent of the executive. Thus, subfederal contacts could conceivably reach the federal level through the cooperation of friendly subfederal units on the other side of the border.

The Trail Smelter Case

The Trail smelter case is often cited in reference to acid rain because of its dictum that: "No state has a right to use or permit the use of its territory in such a manner as to cause injury by fumes in or to the territory of another . . . when the case is of serious consequence and the injury is established by clear and convincing evidence."[98]

The case arose out of the operation by the Consolidated Mining and Smelting Company of Canada, Ltd., of a smelter in Trail, British Columbia, about eleven miles from the U.S. border. Sulfur dioxide drifted downwind from the smelter operations, causing crop damage in Washington State.[99] The dispute developed when the injured Washington farmers decided that the cash payments some had received from the company were insufficient, and they banded together to seek greater compensation.

The smelter company was unable to resolve the dispute by purchasing easements on the damaged property, as it had in Canada, because of a Washington State constitutional provision forbidding foreign ownership of interests in land. Nor were the farmers able to get satisfaction in the Canadian courts, because it was impractical to bring the hundreds of suits in a foreign forum and because the farmers were certain to lose due to jurisdictional provisions of Canadian law. They certainly did not have the resources to carry an appeal to the Privy Council in England in hope of changing the jurisdictional rules.[100]

By 1928 the federal governments of both the United States and Canada had become involved, and they referred the matter to the International Joint Commission, despite the essentially private nature of the dispute.[101] The IJC made an exhaustive investigation of the problem, involving field surveys, hearings, arguments by counsel, and the filing of documentary evidence. In a unanimous report the IJC recommended that the smelter pay $350,000 for injuries up to 1931. This report was accepted by the Canadians but rejected by the United States because it did not address the issue of remedial action. The IJC had split on national lines on remedial measures and hence had not addressed them in the report.[102]

Following the rejection of the IJC recommendations, an arbitral tribunal was eventually set up and an agreement was reached. The tribunal was unusual in several respects. First, there was no dispute between the governments, only one between citizens. Second, the usual rule of international law, that domestic remedies must be exhausted, had not been met because no lawsuit was brought in Canada by the farmers (because of the virtual certainty of loss).[103] These problems were solved by what has been referred to as a "transmutation" of the dispute into intergovernmental claims. In effect, this was merely a waiver by the Canadians of certain procedural rights upon which they could have insisted, and not the major breakthrough in international law it is often claimed to be.[104]

The tribunal eventually awarded the same damage package as that awarded by the IJC, plus $78,000 for injuries through 1937, and required a $20 million control regime.[105] The arbitration ultimately succeeded because there was a strong desire on both sides to settle the dispute.[106] Perhaps the real key was that both governments were interested in solving the resulting political problems rather than in solving the problem per se.

Although the Trail smelter case is interesting from a theoretical point of view, it does not have a great deal to offer to the solution of the acid rain problem. One of the most significant aspects of the case is that it has not been repeated.[107] The reasons are varied, but include: time to resolve the case (thirteen years); lack of interest by either side in allowing a large number of claims; strong interests in protecting industry on both sides of the border during times of high unemployment; and the difficulty of obtaining the close cooperation between federal governments that existed during periods of the Trail smelter case.[108]

In some ways the Trail smelter dispute resembles the acid rain dispute: there were disagreements over the source of the injury and the extent (and existence) of damages, and the issue became political. The major difference, however, was that the dispute was fundamentally local rather than national, and hence involved only one plant rather than an entire industry. An essentially local dispute was transformed into a federal concern on both sides of the border by continuous local political pressure on Congress and the executive branch. Unlike the acid rain debate, however, there was no counterpart of the polluters on the U.S. side, nor was there a concerned group of victims on the Canadian side.[109] Thus, each government faced pressure only from a single direction. In the case of acid rain, each government is pressured by both victims and polluters.

Summary

At the federal level informal means of pressuring policymakers offer excellent opportunities to other governments. The exploitation of these methods, however, depends on a willingness to listen on the part of the policymakers lobbied. If they have made up their minds before pressure is applied, they are likely to be unaffected. For this reason informal methods are best used at the early stages of a bilateral problem, before too many domestic political considerations have arisen. Informal methods are most promising in dealing with agencies that the lobbying nation can consider allies. For example, Ohio would have a greater chance of success in lobbying Ontario Hydro than in lobbying Environment Canada on acid rain.

The use of these avenues does not promise quick solutions for the

acid rain problem. An example of the time frame involved is the thirteen years it took the United States and Mexico to reach an agreement
on salinity of the Colorado River, despite constant pressure through
the U.S.-Mexico IPG.[110] It is difficult to say whether the group actually
assisted the treaty's formation. However, the Mansfield study claims it
as a success and it seems reasonable to accept the claim.

The benefits of the IPG are extensive, if indirect. The U.S. delegation
summed up the benefits of the Canada–United States Interparliamentary Group in quite general terms: "All delegates agreed that the frank
and friendly discussion of the issues by the committee deepened their
individual understanding of each problem and that such discussions
were useful in reducing friction and promoting mutually acceptable
solutions to problems."[111]

Moreover, these benefits are inexpensive: the sixteenth meeting of
the IPG cost the United States only $1,600. Informal means of pressur-
ing policymakers therefore offer an opportunity for influence on both
sides of the border at low cost.

Subfederal Relations

Because the conflict over acid rain is first a regional one, relations
between the affected subfederal jurisdictions are important. Cooperation between victims or polluters on both sides of the federal borders
would strengthen the coalition's hand in both federal systems. Even
more important, acid rain is affected by a host of small decisions on
energy, environmental protection, and natural resources development.
Close contact between subfederal units can help to recognize the linkage of these decisions to acid rain and can suggest means to resolve
the problems before controls are imposed and resources invested.

This section explores the existing subfederal relations and the
potential for influencing policy development by these means. Particular attention is paid to four cases involving transboundary pollution
—the Atikokan, Cabin Creek, and Poplar River energy projects and
the Michigan-Ontario air pollution cooperation.

Nature and Extent of Relations

Federal, state, and municipal levels of government are involved with
pollution control. Subfederal involvement in transboundary air pollution control is growing rapidly. Subfederal activities are not limited,
however, to cooperation with federal authorities. In many instances
provincial and state governments have taken the initiative and dealt

Table 9-1.   State-Provincial Interactions through 1974

State	Total	Environmental	State	Total
Washington	33	2	Oregon	16
Montana	31	3	Wyoming	3
North Dakota	21	3	South Dakota	5
New Hampshire	25	1	Rhode Island	6
Minnesota	47	13	Iowa	12
Vermont	31	2	Connecticut	10
Illinois	14	7	Pennsylvania	10
New Jersey	14	6	Michigan	56
New York	48	6	Massachusetts	24

Source: Roger Frank Swanson, *Intergovernmental Perspectives on the Canada-U.S. Relationship* (New York: New York University Press, 1978).

directly with each other to resolve problems, even, in a few cases, signing formal agreements. The volume of these interactions is summarized in table 9-1. As far as the legal aspects of such agreements are concerned, U.S. states are more restricted than Canadian provinces in handling foreign jurisdictions. Even with the greater restrictions on formal agreements, states have managed a high degree of independent dealings, primarily on practical issues such as auto registration and adoption reciprocity. State-provincial interaction has not led to major issues in U.S. federal-state relations.[112] Some federal laws, such as the International Bridge Act of 1972, even specifically provide for such agreements.[113] They have been more controversial on the Canadian side, but only because of a few instances where provinces have attempted to establish contacts with foreign national governments (e.g., Quebec and France, and British Columbia and the U.S. federal government).

Five basic mechanisms exist for interactions: (1) ad hoc bureaucratic meetings; (2) an organized presence, such as an official office, in the other jurisdiction;[114] (3) a joint organization, usually a joint committee; (4) a summit between premiers and governors;[115] and (5) legislative exchanges.[116] Table 9-2 illustrates the various interactions.

The bulk of the interactions occur at the bureaucratic level and are not significant in terms of constitutional questions of distribution of power between subfederal and federal levels of government.[117] State-provincial interactions can be classified into three basic types: formal agreements (accounting for 5.7 percent of all interactions classified in the 1974 study for the U.S. State Department), understandings (23.6 percent), and arrangements (70.6 percent).[118] The ministerial and premier/governor levels are involved in the more controversial inter-

Environmental	State	Total	Environmental
1	Idaho	7	0
0	Maryland	12	0
0	Nebraska	13	0
0	Alaska	11	3
0	Ohio	6	3
0	Wisconsin	37	5
4	Maine	110	4
5	Indiana	8	3
0	Delaware	4	0

actions, as would be expected. Also, contiguity has been an important factor in the extent of environmental interactions, as can be seen from table 9-3.

Several policy observations are in order at this point. First, the volume of state-province interactions is large and covers a wide range of subjects. Second, most of these relations are of a "maintenance" nature and arise from "the need to coordinate activities between local governments."[119] These activities, while of minor constitutional significance, are important here because they serve to establish an infrastructure of cooperation between the governments and because they show that routine information sharing is not only possible but easily accomplished. Third, and most important, those interactions which do involve, directly or indirectly, challenges to federal authorities, share a number of characteristics: (1) the perceived importance of the issue is high for the subfederal unit; (2) the jurisdiction feels comfortable constitutionally with its actions, either because of its own interpretation of its prerogatives (as in Canada) or because the action falls short of a direct constitutional challenge (e.g., the Michigan-Ontario joint communiqué criticizing the Nixon administration's impoundment of water pollution control funds); and (3) the subfederal jurisdiction feels that insufficient attention is being paid to the issue or at least to its particular interests at the federal level.[120] Before proceeding further, some examples of the specifics of relevant interactions will be presented.

Table 9-2.　A Conceptual Framework for Provincial-State Relations

Bureaucratic	Ministerial	Premier-Governor
Professional network contacts	Goodwill visits	Goodwill visits
information exchanges	Consultation with local governments	Consultation with local governments
	informal	informal
Consultation with local governments	formal	formal
policy comparison	Treaty negotiation	Treaty negotiation
problem resolution	sanctioned	sanctioned
membership on joint bodies	unsanctioned	unsanctioned
	Treaty establishment	Treaty establishment
Operation of offices abroad	sanctioned	sanctioned
	unsanctioned	unsanctioned
Delegated treaty servicing		
Treaty negotiation		
sanctioned		
unsanctioned		
Treaty establishment		
sanctioned		
unsanctioned		

Source: P. R. Johannson, "British Columbia's Relations with the United States,"
Canadian Public Administration #21 (1978): 225, figure 4.

Examples of Subfederal Relations

The immersion of the states and provinces in the acid rain debate comes in large part from the two countries' widely diverging views of the problem. Both nations are polluters and receptors, and subfederal interests often conflict. For example, Ohio is not interested in reducing its emissions, while both Ontario and New York desire it to do so. In the United States the major receptor states are the New England states, New York, Michigan, Wisconsin, and Minnesota. (Both New York and Michigan are also significant polluters.) Ohio, Illinois, Indiana, and West Virginia are the primary polluters.[121] In Canada, Ontario and Quebec are both polluters and receptors, while the Atlantic provinces are primarily receptors. The western provinces are primarily polluters.

Maine and New Brunswick have formed a task force with the two federal governments to clean up the St. John River basin. Quebec and Vermont have cooperated in lake cleanup.[122] British Columbia and Washington State signed a formal agreement on cleanup of oil spills

Table 9-3. State-Province Interactions, by Geographic Area

Area	Percent	Average number per state
New England	36	31
Midwest	31	20
Border	62	—

Source: Roger Frank Swanson, *Intergovernmental Perspectives on the Canada-U.S. Relationship* (New York: New York University Press, 1978), p. 236.

several years before a federal-level agreement was reached.[123] They also established a joint legislative committee to coordinate pollution control with respect to oil tankers.[124] Many instances of small-scale cooperation on environmental issues have been reported. One example occurred in Vermont when Derby, a small town near the border, sought EPA funding for a Canadian sewage treatment plant's expansion to handle the town's sewage. Prospects for the funding were reported to be good.[125] The most active province generally, and particularly with respect to acid rain, has been Ontario. Its actions have ranged from lobbying Walter Cronkite at his provincial summer home to tours for state legislators and journalists.[126] Provincial representatives have testified at U.S. congressional hearings, and Ontario takes at least partial credit for helping to engineer the Memorandum of Intent on Transboundary Air Pollution.[127]

Four cases are examined in greater detail below. These cases were chosen both because they illustrate the flow of pollution in both directions and because they involved state/provincial governments. They are relevant to acid rain because they show the successes and failures of that involvement.

Michigan/Ontario   On the Canadian side, Ontario actively denounced air pollution imported from Michigan. The minister of the environment appeared before the State of Michigan Air Pollution Control Commission in a hearing over Detroit Edison's request to delay compliance with state emissions standards.[128] Ontario's presentation invoked the Trail smelter case and the Memorandum of Intent in support of its request that the board take into account the effect of the delay on Ontario. It also outlined a detailed critique of the power company's cost-benefit formula and discussed acid rain at length.

Michigan and Ontario had previously cooperated extensively on air pollution in the Detroit-Windsor area. Following the August 1971

Governors' and Premiers' Great Lakes Conference, a committee was established with the heads of the air pollution agencies for the state of Michigan, Wayne County (Michigan), and the province of Ontario. This committee developed "an integrated cooperative program for the abatement of transboundary air pollution."[129] No formal international federal agreement was ever achieved because of U.S. federal resistance.[130] Yet provincial-state cooperation was extensive enough to include a joint communiqué issued in April 1974 criticizing the U.S. federal government for reducing air pollution control funding.[131]

In 1974 the premier of Ontario and the governor of Michigan signed a memorandum of understanding that set up the Michigan-Ontario Transboundary Air Pollution Commission (MOTAPC) to exchange information and data and to compare results. The International Michigan-Ontario Air Pollution Board was also created, to monitor MOTAPC and report on it to the IJC and hence the federal governments. One example of the achievements of MOTAPC was an agreement between Wayne County and utilities companies operating within it to adjust operations in accordance with changes in the Ontario air pollution index.[132] This cooperation was not based on any legal requirement, but simply on "jawboning."

Cabin Creek Coal Project   Montana and British Columbia became involved in a dispute over the development of a coal mine in British Columbia that threatened the Flathead River area of Montana. The project involved several open-pit mines covering an area over one mile wide and 2,400 feet long.[133] Pressures against the project developed on four fronts. First, a coalition of environmentalists and sportsmen on both sides of the border attacked the project.[134] Second, Montana put pressure on the State Department to influence the Canadians.[135] Third, Congressman Max Baucus, who had made the project a major issue in his election campaign, worked through the State Department and also met directly with project officials. Finally, Montana passed a statute enabling state courts to issue injunctions against foreign nationals and another authorizing a study of the effects of the project on Montana.[136] The project was eventually discontinued for economic reasons. This case illustrates both how subfederal units can get involved with other jurisdictions, and the need to get involved early, before crucial policy decisions are made.

Atikokan Power Project   The Atikokan project was a 1976 proposal by Ontario Hydro, a Crown corporation, to build four 200 Mw coal-fired generating units approximately twelve miles from Quetico Provincial Park and thirty-eight miles from the U.S. border and the north-ern edge of the pristine Boundary Waters Canoe Area.[137] Negotiations began on the federal-federal level in 1977 with a Canadian agreement to provide the results of studies to the United States. However, in early 1978 the Canadians turned down a request for scrubbers on the plants and for a reference of the matter to the IJC. Eventually the two coun-tries decided on a joint research program.[138] Finally, in part because of pressure from the United States, the Canadians revised the plant plans and settled on smaller plants to be used for peak power rather than baseload generation, and with only 40 percent of the original capacity. The revisions reduced to one-eighth the original SO_2 emis-sions estimates.[139]

This dispute, and the Poplar River dispute (see below) led to the Foreign Relations Authorization Act of 1978, with its mandate for nego-tiations between the United States and Canada on transboundary pollution.[140] The Canadians rejected any reference to the IJC of the matters because they did not admit that there was a potential for harm to the United States.[141]

Poplar River   In March 1972 the Saskatchewan Power Corporation (SPC), a Crown corporation, applied to the provincial government for permission to store water from the East Fork River. In July 1972 the province reserved for SPC six thousand acre-feet annually for five years. Soon thereafter the SPC announced plans for a coal plant, lignite mine, and reservoir along the East Fork. In early 1975 the IJC began an inves-tigation of the projects' air and water pollution and water allocation consequences. The day after the IJC announced it would investigate, the province authorized the SPC to begin construction. Two months later the Canadian federal government issued a five-year license for the dam and reservoir, which was renewed in 1980 for an additional thirty-five years.[142]

From the beginning the plans for the plant were a subject of contro-versy between the United States and Canada. Nevertheless, Saskatche-wan continued to develop the plant.[143] However, both the U.S. federal government and the Montana state government met several times with the Saskatchewan and Canadian governments to review the plans for the project. EPA undertook a study of the issues and concluded that the plant would violate neither the Montana nor the U.S. federal SO_2 limitations.[144] Officials from both subfederal jurisdictions participated directly in the federal negotiations, and Saskatchewan modified the plant extensively during construction to meet the IJC's objections.[145]

Several points about Poplar River are of interest here. First, despite the repeated assurances by the Canadian governments that U.S. interests would be taken into account, there was no formal mechanism outside of diplomatic contacts to ensure that U.S. interests would indeed be considered. The only hearings held on the matter where U.S. interests were presented were those conducted by the IJC, and these were held long after the primary issue of whether to build the plant had been decided. (At those hearings both the Montana and the Saskatchewan governments presented information, as did municipalities, citizens, and federal representatives.) Because Poplar River (and Atikokan) are considered strong examples of Canadian-U.S. cooperation,[146] questions arise about the effectiveness of the current level of relations during the early stages of policy decisions. Cooperation took place only after problems had surfaced, and then the two provincial governments were reluctant to change their plans to protect U.S. interests. Without such a willingness on the small scale represented by these projects, it is difficult to imagine U.S. responsiveness on the large scale needed to resolve the acid rain dispute to Canadian satisfaction.

Second, the major joint action taken by the federal governments was the establishment of a monitoring committee to exchange data (as opposed to collecting it). No attempt was made to develop a mechanism, formal or informal, to prevent future disputes—which will surely arise as both the United States and Canada undertake the development of western energy reserves at the expense of the environment. Third, the IJC found its fact-finding role in this case "enormously difficult" because of the lack of baseline data.[147] Finally, the first recommendation of the IJC was to establish a mechanism to compensate existing water users in Montana for the decrease in quality and quantity of their water.

Summary

Existing subfederal relationships provide an important means of influencing policy across borders. They offer numerous opportunities for policymakers on both sides of the border to consult with their counterparts in an effort to reduce tension. The Michigan-Ontario cooperation is an example of success with this approach. Poplar River and Atikokan were not complete successes, however, and point out some of the problems with subfederal relations. In both cases U.S. opposition did not meet with great success in altering the projects. Economic considerations forced eventual changes in the projects, but little of the change can be attributed to Montana's efforts.

Several differences between the two groups of cases can be identified. In the successful case, pollution was a problem on both sides of the border, as with acid rain. Moreover, both governments were strongly motivated to work together to reduce the problem. In the second group, flow of pollution was primarily one way, toward the United States. Energy and environmental concerns conflicted more sharply than in the other case. Acid rain shares this characteristic. Both Atikokan and Poplar River revealed Canadian insensitivity to U.S. concerns. Increased cooperation of Canadian authorities, particularly at the provincial level, will boost Canada's case for similar treatment in the United States.

Many of the future transboundary problems will lie in the West, while acid rain is an issue primarily in eastern Canada. This fact stands in the way of a solution. The western provinces and states are extremely sensitive about their role in energy development and will not be easily convinced of the need for coordination with their neighbors. Nevertheless, by increasing its own sensitivity to U.S. concerns, Canada can build relationships that will enable it to influence U.S. policy more effectively.

Policy Conclusions

A pragmatic view of the requirements for a successful resolution of any pollution problem shows the importance of subfederal jurisdictions in both designing and implementing solutions. This need has been recognized in the past during the negotiations of the Great Lakes Water Quality Agreements and other environmental agreements. In the case of Canada, there have been both provincial, ministerial-level involvement in the negotiation of treaties and pretreaty provincial-federal agreements that laid the basis for the federal-federal agreement.[148]

Similar factors are true for the U.S. side, where state resistance to a federal plan can delay or frustrate enforcement and implementation both through court challenges and through refusal to cooperate sufficiently to allow the program to function effectively. In both nations the role of subfederal jurisdictions in environmental and resource decisions is growing.[149]

Local officials often make decisions that affect environmental policy. Increased contacts between these actors will make a broader information base available for these decisions. The Ohio Public Utilities Commission, for example, is involved with decisions relating to the expenditures of utility funds for air pollution control and rate structures.[150] Local jurisdictions also often maintain information-gathering networks which could be coordinated and standardized to produce more useful data—for example, of the twelve acid rain monitoring networks in Canada, only five are federally maintained.[151]

Third, many international conflicts, like acid rain, involve technical matters such as pollution control. These issues are qualitatively different from the traditional international issues like military security.[152] Thus, these "new" types of issues impinge on areas which have traditionally been state responsibilities. Domestic politics are also increasingly a factor in these types of issues. As the issues become more political, federal governments are increasingly unable or unwilling to act on them because of the lack of national consensus. Involving the subfederal actors in the policy process restricts the actors' freedom to raise the political stakes with rhetoric from outside the policy process, and forces them to address the problem from a policy perspective rather than from the perspective of attempting short-term political gain.

Finally, increased involvement of subfederal jurisdictions is desirable because of the need to keep informed of the actions being taken by the various actors. State and local officials make many decisions which affect pollution control, and keeping abreast of these decisions is easier if their agencies are involved in the policy process. Ontario, for example, was informed of the SIP relaxation applications of six midwestern states by New York State officials,[153] and consequently was able to intervene in the proceedings. A mechanism whereby the midwestern states could have discussed their plans with Ontario before initiating them would have been preferable. Nevertheless, the example illustrates the efficacy of across-the-border communication.

How can a nation affect policy in another nation to resolve a problem like acid rain? It is apparent that, before the issue can be resolved on the national level, some sort of national consensus on the problem must exist. As yet no such consensus exists in the United States, unless it is a consensus of lack of knowledge. In Canada there is a consensus in the East, but that consensus depends on the western provinces not becoming involved, for they are primarily sources rather than receptors of SO_2 emissions and are committed to development of their energy resources. This will necessarily involve some environmental degradation. To reach the national consensus required for national action, or to maintain it in the face of regional demands for attention to other problems, will consume energy and resources better put to resolving the problem or addressing other societal needs. The time required to reach a consensus is a commodity which may be lacking in the acid rain case as well; by the time consensus is reached, the lakes of New York and Ontario may be dead. A means to reach resolution of the issue directly, bypassing a full national consensus, would save time and money. As a means of identifying useful mechanisms for affecting policy development, we consider two hypothetical government structures.

The One-Nation Approach

Try to imagine a situation in which the national boundary between Canada and the United States does not exist. The acid rain dispute becomes a dispute between polluters and victims rather than between two sets of regions competing both within national boundaries and across them, and the problem is much simplified. This is not as unrealistic as it at first seems. The two nations share a common airshed, common resources, common problems, and to a certain extent a common economy. The boundaries between the two nations are eighteenth-century administrative ones, not natural borders. The systems of law are similar; if the details vary, they do so by no more than the laws of Louisiana and Texas or Quebec and Ontario, and less than the legal systems of the European Community.[154] Many corporations and individuals are accustomed to freely operating simultaneously in both systems. So far, the conflict over acid rain is a regional one, and one between regions that are not completely different. Thus, New York and Ontario have enough in common to participate in a policy debate between the victims and the polluters, as do Ohio and Ontario or Quebec.

Moreover, given the import of local decisions on acid rain (e.g., whether to grant industrial development bonds to a smelter), the involvement of local decision makers is critical. Magic labeled "Section 115" or even "Clean Air Act" is insufficient to resolve such a complex problem. Decision making over a wide range of areas needs to be integrated. Failing that, a perhaps unrealistic goal in light of nationalist feelings, decisions should at least be made with awareness of their consequences.

The European Approach

What if North America were many nations instead of two? Imagining a European version of North America points to several interesting possible solutions. Two groups would most likely band together in a common forum—the polluters and the victims, as has happened in

Europe. Efforts would be made to cajole and harass polluters into abatement and to quiet victims.

Returning to reality, we see that our North American Europe has a great advantage over the real Europe. Here we have two federal structures that can be used to prod the polluters into taking remedial action. Thus New York could, in theory, convince the federal government to stop Ohio from sending SO_2 its way. The Maritimes similarly could secure a federal ban on emissions from INCO. Neither of these two events is at all probable. However, New York can, and has, taken advantage of the federal courts and federal legislation to encourage Ohio to reduce emissions. Federal money can alleviate the burden of abatement costs, and federal research can lead the way in new abatement technologies. Thus the federal role is an important one. But it is the subfederal jurisdictions' interactions with each other that will shape the solution to the acid rain problem. Naturally, neither the one-nation approach nor the European approach is realistic—but they provide several policy lessons. Since a full bilateral federal agreement is unlikely, the most effective means for Canada to influence U.S. policy, and vice versa, is for subfederal actors to present their views directly to the policymakers across the border during the policy-making process, much as two members of the same federal structure might.

Summary

The tools and structures examined in this chapter are effective ones for presenting friendly neighbors' policy views. Private lawsuits like *Michie* can provide publicity and in some cases a measure of relief from pollution. Interparliamentary contacts at all levels provide opportunities for informal discussion of policy between legislators representing regional interests. Intervention in administrative proceedings may force consideration of alternative policies—as it did in the Alaska pipeline case. Section 115 and Bill C-51 proceedings offer a direct opportunity to address foreign impacts of pollution control decisions. An advantage of almost all these methods is that little or no federal action is needed to initiate them. This reduces the time frame for initial action.

There are, of course, problems with this type of approach. A major reason we have federal governments is to prevent states or provinces from behaving as if they were independent. However, here we are calling for free interaction, not independence. Federal oversight and judicial review should prove sufficient to keep the interactions within reasonable bounds. The IJC oversight of the Michigan-Ontario air pollution cooperation, with reports to both federal governments, is a good example of this.

States and provinces have few formal bargaining chips to offer each other in exchange for SO_2 reduction.[155] But by getting policymakers together for discussion, we can increase the chances of a mutually acceptable resolution of the problem. Hence an informal regional body modeled on the IPG, without minutes or pressures for a formal agreement or other document, should be created to foster discussion on issues that affect the common airshed. (Although the IPG itself is not a resounding success, its problems stem from its perceived lack of prestige, not from any inherent structural defect.)

The group would probably be small at first, with only the victim jurisdictions participating, but it could eventually grow into a larger body, even including the major polluters, once they were convinced that the group was not to be a forum for speechmaking. Such an organization would require a more formal counterpart to tackle issues such as coordination of data collection, perhaps one modeled on the Bilateral Research Consultation Group or the IJC. The separation of the two functions would preserve the advantages of the IPG type of structure without forcing the body to grapple with management problems for which it is not equipped. Establishment of such a second organization would have to be a second phase after the development of the informal group because it would require much broader participation to operate successfully.

Major agreements at any level appear to be unattainable at this stage. What is needed is an opening of the policy process at all levels to input of affected parties. A regional airshed group is one way to achieve some measure of this. Equal-access reforms are another. They and the other means of achieving these ends provide increased communication and hence increased opportunities for resolution. None of these measures is a substitute for direct action by the two federal governments. However, each provides a means of both nudging the other nation closer to an agreement and affecting current policy development. The two functions have very different time frames. For the latter, the effects range from the present to several years in the future. For the former, the time needed is obviously much greater. As actions move along a continuum toward formality, the time necessary for each grows longer.

Reaping the immediate benefits of a lawsuit merely requires filing it and announcing that to the media. Temporary injunctive relief may also be immediate. To benefit from a permanent injunction or damages requires years of litigation, and to influence policy significantly may require many lawsuits. Thus each action yields a range of benefits at different times.

The most important lesson from this analysis is not the specific proposals, which are provided as illustrative examples only, but the identification of several broad areas where action can take place to influence policy. These actions are important because they offer steps which can be taken individually as well as jointly. Without any cooperation from the United States, Canada could undertake to reduce the substance and procedural difficulties between the two nations, thus helping to close the rhetoric gap that U.S. critics of Canadian acid rain pronouncements can point to. The United States, on its side, could take Canadian concerns seriously and encourage subfederal cooperation.

Chapter Ten
Conclusions

||||||||

We have followed acid rain policy from the time when it first appeared on the agendas of Canada and the United States in the mid-1970s to the final months of the Reagan administration. Acid rain presents a much more complex environmental policy issue than those previously addressed. From a political perspective its complexity derives from the fact that two sovereign nations, one of which sees itself as the victim of the other's air pollution, are involved. Scientifically it is complex because much remains to be learned about the long-range transport of air pollutants, the establishment of source-receptor relationships, and the impacts on ecosystems.

In drawing some conclusions, we return to our original questions concerning the success that each country has had, both individually and jointly, in addressing the problem of acid rain. Although the main focus of this book has been on the development of a bilateral agreement for the control of acid rain, we acknowledged early on that bilateral negotiations would not be successful without a sound underlying domestic policy in each country. Since the first edition of this book was published the Canadians have taken action on the control of acid precursors and have unilaterally implemented the policies that they originally hoped would be enacted jointly. Thus, much of the focus of the debate now rests with domestic policy development in the United States. Because the administration is unable or unwilling to move ahead with the issue, reauthorization of the Clean Air Act provides the best opportunity for acid rain precursor controls. The issues

---

This chapter was written by Jurgen Schmandt and Judith Clarkson.

determining the outcome of this debate are summarized below.

In the Preface we defined three questions that guided our research. We return to them now to summarize our conclusions concerning the state of the acid rain debate at the bilateral level.

Question 1: What is the extent of agreement and disagreement between Canada and the United States ... about the nature of acid rain and its environmental effects?

Many competent science-assessment documents of acid rain and its effects have been prepared. As a result, policymakers have received detailed scientific information on which to build an acid rain policy. The degree of disagreement between scientific reports from the United States and Canada is small; it is larger among experts drawn from different institutions, such as universities, governments, and industry. Important areas of uncertainty remain, such as human health effects associated with ambient levels of acidic aerosols on a regional basis and the contributions of individual sources of emissions to deposition at specific receptors. Still, we know more about acid rain and its effects than we did about earlier environmental issues at the time legislation was enacted.

Science assessments have become an important part of policy development for technically complex policy issues. Study of the acid rain case leads us to several conclusions about the role of science assessments.

On the positive side, the science-assessment process did what it is supposed to do: review, summarize, and interpret scientific knowledge in order to provide an empirical base for decision making. Specifically, science assessments on acid rain provide two important decision aids:

1. Authoritative guidance on research hypotheses and findings: The assessments supply the policymaker with information on the nature of the acid rain phenomenon, its probable causes, and its effects. They also tell him whether we know enough to qualify available knowledge as preliminary, contradictory, or firm; what gaps in knowledge exist and how they might be closed; and whether conflicting evidence is of critical importance or not.

2. Summation of complex research findings in nontechnical language: The scientific experts involved in preparing and reviewing assessment documents have provided a scientific synthesis that is accessible to the decision maker without losing its accuracy.

The acid rain case also illustrates several problems with preparing and using science assessments. First, the desire for complete information can easily be used as part of a delaying strategy. Such delays may well be less the result of asking for more scientific assessments than of conducting more research.

A second problem concerns the limits of scientific advice. The science assessments that we examined did not go beyond a fact-finding role. Specifically, they avoided prescribing solutions or making policy judgments. However, the borderline between neutral assessment and policy advocacy is difficult to draw. The leading independent scientists who have studied acid rain conclude that the issue is urgent and serious. Some go further and warn that damage will be irreversible. A cautious policymaker can read bias into these conclusions. This is perhaps one reason why government-employed scientists, rather than independent scientists, were asked for advice in support of bilateral negotiations. The NAPAP interim assessment was compiled almost exclusively by government scientists. The U.S. government made little use of the National Academy of Sciences, which has long played a leading role in providing the executive branch with advice relating to science policy. Nor was the International Joint Commission for the Great Lakes involved, even though this small organization has a proven record for providing reliable scientific information to the two governments.

A third problem may be a direct consequence of the second. Many assessments did not address certain critical areas. Could it be that the scientists involved in the preparation of assessment documents were overly cautious in defining the scope of their inquiry? It is difficult to make this charge because experts can rejoin that not enough was known to address critical issues. Even so, a more prestigious group of advisers might have done a better job of putting into perspective some of the least understood but possibly most serious dangers associated with acid rain, such as soil toxicity, forest damage, health effects, and reversibility of damage.

A last problem with the analytical effort concerns the imbalance between scientific and policy assessments. Compared to the massive effort in generating and synthesizing scientific information about acid rain, less effort was spent on preparing reliable estimates of the costs of damage and control options. Initially, the major studies in the United States came from congressional agencies. However, with the formation of NAPAP, a program specifically initiated to investigate the scientific and policy aspects of the acid rain debate, more emphasis on the social and economic considerations might have been expected.

In its interim assessment NAPAP claims that a lack of quantifiable effects constrained its ability to evaluate the economic impacts of acid deposition.[1]

Even less has been done to study legal and policy options. The difference is striking when one compares the documents that were prepared by the bilateral work groups on transboundary air pollution.[2] Work Groups 1 and 2 prepared detailed reports on the science of acid rain and its environmental impacts. By contrast, the work group investigating emissions, costs, and engineering limited itself to an industry-by-industry assessment of emissions, control technology, and cost estimates. No comprehensive analysis of control costs was presented.[3] The work group in charge of legal and institutional analyses was even more cautious, and produced only an interim report describing the applicable legal provisions in both countries.[4] No analysis of policy options was ever published, and the work group assigned to investigate this area was dissolved prematurely.

We conclude that governments have become accustomed to the need for detailed science assessments but are still uncomfortable with conducting or sponsoring economic and policy analyses of emerging issues. Or if they prepare such studies, they may not publish them. The Reagan administration certainly opposed planning for a control policy that it did not want. When William Ruckelshaus at the EPA departed from this norm, the plan that he presented was easily defeated by more influential administration officials. Given this situation at the governmental level, we were surprised to find little effort on the part of universities and independent research institutes to fill the void and contribute to policy development with economic, legal, and policy analyses of the acid rain issue.

Question 2: What domestic policy developments are under way in Canada and the United States aimed at controlling acid rain, and what additional measures would be helpful?

We remain convinced that national policy development for acid rain will have to take precedence over bilateral action. Since 1984 Canada has made progress in controlling the emissions of acid precursors. It has enacted federal legislation to bring automobile emission standards into compliance with those in the United States. In addition, it has negotiated emissions reduction targets from stationary sources with each of the seven eastern provinces, most of whom have now implemented control programs. The combined efforts of these seven provinces will result in a 50 percent reduction in SO_2 emissions. The overall goal is to reduce acid deposition to eighteen pounds of sulfate per acre per year. In order to realize this, emissions from U.S. sources will also have to be reduced by 50 percent.

Legislative proposals to control acid rain are currently being considered as amendments to the Clean Air Act, which is due to expire in August 1988. Reauthorization has been in progress since 1982, and several extensions have been negotiated since that time. Attempts to control SO_2 and NO_x emissions have provided some of the main sources of controversy. Because so much is at stake for the utility and coal industries, there has been heavy pressure from opponents of emissions controls. During the course of the last eight years, programs to control emissions have taken a variety of forms in an attempt to arrive at a compromise. Initial attempts were focused on the states east of the Mississippi. Regional controls represent a radical departure from the uniform national standards that have been the mainstay of environmental policy since the late sixties. However, increasingly there is evidence that the problem is of national scope, and this may determine the outcome of the debate, despite the political trade-offs associated with extending controls to western states.

Policy development for acid rain can be divided into three stages: issue definition, policy design, and political mobilization. Most activity so far has focused on the first two. The third stage has been stalled almost from the onset of bilateral negotiations in 1978. The pressure for and intensity of debate about policy mobilization have changed several times during this period. Canadian efforts to arouse public support for acid rain controls in the United States were designed to overcome the stalemate. It seems that this strategy has failed. The Progressive Conservative government elected in 1984 has called off public intervention in U.S. policy development for acid rain, opting instead for a return to quiet diplomacy. In the United States acid rain did not become a major issue in the presidential campaign of 1984 and so far has not emerged as an issue in the 1988 campaign. The Reagan administration remains firmly opposed to additional controls, while supporting additional research, demonstration projects, and monitoring.

The role of political mobilization in policy implementation and the factors that influence it are shown in figure 10-1. These factors all lend themselves to analysis within the context of acid rain policy formulation in the United States. They form the basis for the discussion relating to the enactment of legislation for greater emissions control standards.

Figure 10-1.   The Relationship between Political
Mobilization and Other Factors.

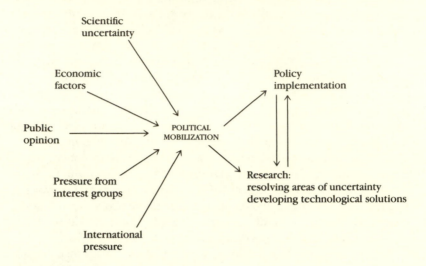

Legislative Attempts to Control Acid Rain

Scientific Uncertainty   This argument seems to have almost run its course. Although the exact relationship between sources and receptors is still unclear, the scientific community in general agrees that there is no doubt that a causal relationship between SO_2 and NO_x emissions and acid deposition can be established at the regional level. There is no evidence that the need for controls is diminishing. Because source-receptor relationships are not well known, the largest area of uncertainty relates to the effectiveness of controls. It is not known precisely what level of controls will realize what benefits. However, this is hardly an argument for no action.

Economic Factors   The economic arguments must be taken seriously.[5] On the other hand, it is all too easy for the industrial groups that are most directly affected to exaggerate the costs of controls while ignoring the costs of inaction. For instance, it has been calculated that in Canada acid deposition threatens industries that account for 8 percent of its gross national product. Much of the difficulty in estimating costs and benefits derives from the issue of uncertainty. However, in addition to this problem is the added difficulty of comparing cost estimates in which different assumptions were made. Examples alluded to in chapter 5 include differences in the changing rate of utilization of various plants, differences in annualization rates for control costs, and differences in projections for low-sulfur coal cost premiums.

Public Opinion   At present there is not enough support for successful policy mobilization in the United States. In part this may be due to the regional character of the acid rain threat. As long as large parts of the country believe that they are unaffected, they will not actively support costly controls. But even in the affected regions the perception of damage is not serious enough to overcome opposition to new taxes or fear of lost jobs. Significant policy development, therefore, is not likely to occur until research, media, and political activists can demonstrate to large numbers of people that serious danger is at hand. The example of West Germany illustrates how dramatically a country's stand on the issue can change. For years Germany had opposed controls, but when damage to forests was reported the public became aroused and control measures were suddenly politically acceptable. In summary, it is fair to say that the majority of Americans (more than 80 percent in most opinion polls) support acid rain controls, but, given the range of problems facing the country, it is not very high on their list of priorities.

Interest Groups   Probably the largest single factor contributing to the apparent inertia in the acid rain debate has been the role of interest groups. The utility and coal industries have spent millions of dollars to prevent enactment of legislation requiring stricter emissions controls. The availability of such resources results in a disproportionate amount of influence on the political process by a few select groups, which are able largely to counteract the impact of public opinion. These groups have exploited the issue of scientific uncertainty, claiming that the enormous expenditures that would be required cannot be justified in light of the highly speculative nature of the benefits that might result. They claim that the Clean Air Act in its present form provides adequate protection and that the best strategy is to wait for the development of more advanced clean-coal technologies.

This view is shared by the Reagan administration. Lee Thomas, administrator of the EPA, is one of the most effective lobbyists that industry groups have. He often participates in Senate committee hearings and is opposed to any provisions that strengthen the Clean Air

Act. Despite the intransigent attitude of the administration, the mood in Congress is more conciliatory and probably more closely reflects the feeling of the country. When representatives of groups on opposing sides of the issue are given the opportunity to engage in constructive dialogue, each side is usually willing to concede on some points, and broad areas of agreement are often possible. This point is illustrated by the Acid Rain Partnership Project, a year-long effort by citizens of Ohio and New Hampshire to develop consensus recommendations for the control of acid rain.[6]

International Pressure   Canada has been trying unsuccessfully for almost a decade to get the U.S. administration to agree to acid deposition controls. In bilateral negotiations they have repeatedly tried to get an agreement on a timetable for specific reduction targets. In one strategy in the early 1980s the Canadians tried to arose public sympathy by using techniques that the U.S. government found offensive. With the election of a more conservative administration in Canada, the emphasis in the United States has been to focus more on educational and less on political activities. Because the economy of Canada is so much smaller than that of the United States, it is very difficult for the Canadian government to apply pressure. In general, linkage to other issues is considered undesirable, and given the disproportionate influence that Canada and the United States have on each other's economies, it is unlikely that Canada could have much effect. In addition, the bilateral U.S.-Canadian relationship is more important to Canada. A much larger percentage of Canadians live near the U.S.-Canadian border than do U.S. citizens. Thus, Canadians are not only more aware, but are also more heavily influenced by U.S. actions than the other way around.

Policy Implementation   With such strong opposition both from industry and the administration, it is not surprising that reauthorization of the Clean Air Act and the inclusion of acid deposition controls have met with difficulties. Although there is considerable support in Congress for such action, several powerful members, including Senate Majority Leader Robert Byrd, are strongly opposed. Senator Byrd, who represents West Virginia, a state heavily dependent upon high-sulfur coal, is in a position to prevent such legislation from reaching the floor of the Senate. Much effort has been invested in trying to develop compromise legislation that will accommodate, at least to some extent, the concerns of opposing constituents. Over the course of time, more attention has been paid to the issue of equity by extending controls nationwide and limiting residential utility rate increases. In addition, in the interest of economic efficiency, there has been less emphasis on specific control measures and more on overall reduction targets.

Research   Research can be a valid part of the policy formulation process if it is designed to answer specific questions that relate directly to the need for and the effectiveness of control strategies. As mentioned earlier, the emphasis has been on resolving the scientific uncertainties rather than concentrating on the costs and implications of various control options. As such, it would appear that research is being used as a delay tactic. More recently, the Reagan administration has focused its attention on one long-term potential solution advocated by utilities, clean-coal technology. Although this may be the best, most economical long-term solution, a variety of control technologies are currently available. Environmentalists claim that the Reagan proposal to fund this type of research is merely a means of subsidizing polluters without advancing the cause of acid precursor control. In their view, research should not be seen as an alternative to implementation of existing controls if for no other reason than because "failure" buys more time. In addition, experts from the pollution control industry have testified that federal funding to develop new technologies is not what is needed, but legislation that requires emission controls, thereby ensuring that there is a market for such technology when it is developed.

Opportunities under Existing Law

What about incremental improvements? A partial alternative to new controls for acid rain exists. Existing control programs for the two major sources of acid rain—SO_2 and NO_x—could go a long way toward buying time if they were implemented aggressively. Legally the United States is better prepared than Canada to follow this strategy because the U.S. federal government has the authority to control emissions from new stationary sources and motor vehicles. In Canada this authority has remained at the provincial level.

Paul Sabatier and Daniel Mazmanian have suggested that implementation suffers from natural erosion. According to their model, about five years after enactment of the enabling statute the process of gradual erosion of constituency support begins. The process will "nibble away" at the statute and undermine effective implementation.[7] Implementation of the Clean Air Act follows this pattern only partially. Large

industrial polluters try to extend the lifetime of uncontrolled installations by making use of exemptions made available in the statute. Grad-
ually, new facilities come on-line that control emissions with effective
though expensive scrubbers. Sulfur dioxide emissions from stationary sources have been reduced, suggesting that controls have begun
to make a difference. Automobile exhaust controls are less well
enforced. But explicit policy, rather than gradual erosion of support,
accounts for the lack of more decisive impact of the existing controls.
During both the Carter and Reagan administrations, high-level policy
changes were introduced that had the effect of reducing existing controls. Initially the new emphasis on coal-based generation of electric
power was seen as a necessary response to the energy crisis. Later,
the antiregulation stance of the Reagan administration combined
forces with industrial and union interests in the Midwest to relax
controls.

The Reagan administration, reflecting its opposition to acid rain controls, and to regulation in general, has failed to use existing opportunities within the Clean Air Act. This position has recently been con-
tested in the courts, which, on appeal, sided with the administration.[8]
Not only does the administration claim that existing statutes do not
provide for the control of acid deposition, but they feel that sufficient
progress in improving air quality is being made. The control of SO_2
emissions under existing statutes has resulted in a decrease in ambient levels of 27 percent between 1973 and 1985. Between 1977 and
1985 SO_2 emissions from coal-fired plants decreased 11 percent,
despite a 40 percent increase in coal use.[9]

*Question 3: What joint measures have the two countries taken to
resolve the issue, and what additional bilateral measures, in light of
the past record of resolving environmental disputes between the two
countries, can reasonably be expected?*

Given our premise about the dominance of domestic policy developments, we never expected bilateral measures to play the leading
role. As we look back, we see no reason to change this assumption.
Three points need to be stressed. First, the bilateral negotiations may
have failed because the agenda established by Washington and Ottawa
downplayed the politics of acid rain. The two governments built an
elaborate mechanism of work groups to prepare for diplomatic negotiations. A strong motivation for creating new, temporary mechanisms
was the desire to keep close control of the fact-finding activities. We
commented earlier on the virtual exclusion of independent research

organizations and the International Joint Commission from the process. Here we go further. Perhaps the negotiations were flawed from the beginning because they tried to reach technical agreements without sufficiently exploring the basic policy issues. From the available record it seems that the negotiations centered on technical fixes, such as the proposed 50 percent reduction in current emissions levels. The hard policy issues were never explored in sufficient depth: What are each country's goals? How can we make them compatible? What does each country have to do to put its house in order? Who will pay how much? What mechanisms do we need to implement policy?

Second, the results of an eventual agreement reached by the two federal governments should not be overestimated. Joint research, monitoring, and coordination of national activities lend themselves to international activities. Beyond that, agreements must reflect policies that are made at the national level. The interdependence of national and bilateral policy developments must be built into the process of negotiations. The Great Lakes Water Quality Agreements illustrate the point. No new international control programs were created, but enough was done nationally to allow a coordinated approach to become possible. And both countries retained the flexibility to introduce different national controls.

Third, the record of environmental relations between Canada and the United States provides ample evidence to define the conditions under which negotiations succeed or fail. Disputes are resolved satisfactorily when the target of agreement is a common resource that needs to be protected or shared. Negotiations break down when the actions of one country are perceived as causing the problem. Acid rain will thus be more difficult to resolve than earlier environmental problems. But recognizing shared interests and avoiding finger pointing seem to be simple but proven strategies for successfully negotiating an acid rain agreement between friendly neighbors.

Appendix A
Detailed Document Responses:
Physical and Chemical Characteristics

IIIIIIII

Table A-1. Historical Trends in Acid Deposition

Question 1.a. *Does there appear to be a historical increase in the trend of quantity or intensity of acid deposition?*

Report	Response	Page	Comments
MOI.2	Don't know	pp. 6.14–15	"The general conclusion is that historical air and precipitation data have sufficient uncertainty in them to make it difficult to draw any conclusions regarding long term temporal trends in either the acidity itself or the precursor emissions. This is not to say that trends do not in fact exist, but if they do, the available data are incomplete and too unreliable to quantify their existence."
NAPAP	Don't know	p. 18	"Precipitation in the eastern parts of the United States and Canada has probably increased in acidity since the industrialization of these regions, but a quantitative assessment cannot be made with the data currently available."
CAD.1	Don't know	p. 47	"The clear increase of nitrate in precipitation and of NO_x and SO_2 emissions certainly suggests that the acidity of precipitation has increased in the last 25 years. However, the historical pH data, measured or calculated, do not allow quantification of an acidity increase."

Table A-1.   (continued)

Report	Response	Page	Comments
DYCO	Not addressed	p. 34	Trends in acidity discussed, but no estimate stated.
NRCC	Don't know	p. 35	"There does not appear to be a significant trend in precipitation acidity, based on samples collected over the past several years in eastern Canada. However, the pH of the precipitation is usually found to be less than 5. As pH is measured on a log scale, relatively large changes in H^+ content at this level are required to gain small changes in pH."
NRC.1	Not addressed		Discussed, but no estimates given.
NRC.2	Don't know	p. 13	"Trends in acid deposition in North America have been difficult to discern, and data with which to assess them are sparse."
ON-HY	No	p. 40	"No statistically significant linear trend was found between acidity data and time since 1955. However, the data indicate that acidity increased significantly between the years 1955/56 and 1972/73. No significant changes in acidity are supported by the post-1972 data."
SCAR	Don't know	p. 67	"Monitoring of the chemistry of precipitation has not, historically, been consistent in North America."

Table A-2. SO₂ Emissions in North America

Question 1.b. *Does there appear to be a historical increase in the trend of quantity or intensity of SO₂ emissions in North America?*

Report	Response	Page	Comments
MOI.2	Don't know	pp. 6.14–15	"The general conclusion is that historical air and precipitation data have sufficient uncertainty to make it difficult to draw any conclusions regarding long term temporal trends in either the acidity itself or the precursor emissions. This is not to say that trends do not in fact exist, but if they do, the available data are incomplete and too unreliable to quantify their existence."
MOI.H	Yes	p. 3	"Our historic trends in sulphur dioxide emissions indicate an increase in Canada from 4.5 million tonnes in 1955 to 4.8 million tonnes in 1980 while in eastern U.S. the 1950 level of 6 million tonnes increased to more than 17 million tonnes in 1978."
NAPAP	Not addressed		
CAD.1	Yes	p. 2.76	"Total emissions [of sulfur dioxide] in the eastern United States doubled from 1950 to 1978."
DYCO	No	p. 14	"Particulate emissions (not solely SO₂) declined nationally by 50 percent between 1970 and 1979, largely due to controls on industrial and utility emissions, decreased burning of solid waste, and reduced coal burning by small sources.... Evidence suggests, however, that local concentrations of particulates in the fine or respirable range are generally increasing due to the dispersal of point sources from urban areas."
NRCC	Yes	p. 34	Yes, for North America taken as a whole. Canada and United States taken separately, however, differ: "Anthropogenic SO₂ emissions in eastern Canada increased from 1955 to 1965, but returned to the 1955 level by 1979. Sulphur dioxide emissions in the eastern U.S. increased [107 percent

Table A-2.   (continued)

Report	Response	Page	Comments
			from] 1955 to 1965, before essentially stabilizing at that level until 1978."
NRC.1	Yes	p. 43	"From 1940 to 1960 the reduction of SO_2 emissions from industrial fuel consumption was balanced by increase of SO_2 emissions from electric utilities. Since 1960 the sharp increase in SO_2 emissions was essentially due to the electric utilities."
NRC.2	Not addressed		
ON-HY	Yes	p. 9	For United States: "The available data . . . has shown that man-made emissions increased significantly [40 percent] between 1940 and 1977." For Canada: "Historical trends for anthropogenic SO_2 emissions in Canada are not available."
SCAR	Yes	p. 35	"Emissions of SO_2 and NO_x in North America have increased greatly since the early 1950s, coincident with population growth, heavier resource utilization, industrial expansion, and the proliferation of the private automobile as a means of transportation."

Table A-3.   NO$_x$ Emissions in North America

Question 1.c. *Does there appear to be a historical increase in the trend of quantity or intensity of NO$_x$ emissions in North America?*

Report	Response	Page	Comments
MOI.2	Don't know	pp. 6.14–15	"The general conclusion is that historical air and precipitation data have sufficient uncertainty in them to make it difficult to draw any conclusions regarding long term temporal trends in either the acidity itself or the precursor emissions. This is not to say that trends do not in fact exist, but if they do, the available data are incomplete and too unreliable to quantify their existence."
NAPAP	Not addressed		
CAD.1	Yes	p. 2.77	"Total emissions [of nitrogen oxides] in the eastern United States increased by a factor of two from 1950 to 1978."
DYCO	Yes	p. 8	Net emissions of NO$_x$ will have increased 26 percent from 1975 to 1985. (Figure interpreted from provided histogram.)
NRCC	Yes	p. 34	"Nitrogen oxides emissions in eastern Canada have increased [150 percent] from 1955 to 1977, whereas in the eastern United States NO$_x$ emissions have increased [200 percent] from 1950 to 1978."
NRC.1	Yes	p. 48	Total United States NO$_x$ emissions have increased 174 percent from 1950 to 1975. (Increase interpreted from figures provided in table.)
NRC.2	Not addressed		
ON-HY	Yes	p. 9	For United States: "The available data . . . has shown that the man-made emissions increased significantly [by a factor of 4] between 1940 and 1977."
	Don't know	p. 8	For Canada: "Historical information of this type is not available for Canada."
SCAR	Yes	p. 35	"Emissions of SO$_2$ and NO$_x$ in North America has increased greatly since the early 1950s, coincident with population

Table A-3.   (continued)

Report	Response	Page	Comments
			growth, heavier resource utilization, industrial expansion, and the proliferation of the private automobile as a means of transportation."

Table A-4.   Sources of SO_2

Question 2.a. *What are the major sources of sulfur dioxide emissions in North America: electric utilities?*

Report	Response*	Page	Comments
MOI.2	No	p. 2.5	For Canada in 1978: Cu-Ni Smelters 38 percent, Power plants 16 percent, Other combustion 20 percent, Others 20 percent.
	Yes	p. 2.10	For United States in 1978: Electric utilities 63 percent, Other sources 37 percent.
NAPAP	Yes	p. 30	"In 1977, total man-made emission of SO_2 in the U.S. was estimated to be ... 81% from fuel combustion in stationary sources. [Of this] electric utilities contributed [75 percent], large industrial plants [19 percent], and commercial and residential heating systems [4 percent]." (Proportions interpreted from chart, p. 32.)
CAD.1	Yes	p. 2.42	"In recent years, electric utilities appear to have contributed to more than half the total sulfur oxide emissions."
DYCO	No	p. 13	"The primary anthropogenic sources of atmospheric particles are stationary fuel combustion [39%] and industrial processes [44%]."
NRCC	Yes	p. 58	"With respect to current emissions of SO_2, thermal power plants contribute approximately 60% of the North American nationwide emission total."
NRC.1	Yes	p. 41	"East of the Rocky Mountains [in the United States] coal [combustion] contributes 71% ... of the total sulfur dioxide emissions.... In 1973, utilities contributed 60% of the SO_2 emissions east of the Mississippi."
NRC.2	Not addressed		
ON-HY	Yes	p. 7	For United States: "The most recent data for 1977 show that electric utilities were responsible for over 60 percent of the total anthropogenic SO_2 emissions in the

Table A-4.   (continued)

Report	Response*	Page	Comments
			United States, while other stationary combustion sources significantly decreased their contribution to about 15 percent."
	No	p. 7	For Canada, 1974: "Industrial processes such as nonferrous smelting accounting for 75 percent of the total emissions and electric utilities being responsible for only 9 percent nationally, and 14 percent in Ontario."
SCAR	Yes	p. 19	Taken together, U.S. and Canadian utilities accounted for 58 percent of anthropogenic emissions.

*Yes = mostly from electric utilities; No = from different sources.

Table A-5.   Sources of NO$_x$

Question 2.b. *What are the major sources of nitrogen oxide emissions in North America: electric utilities, transportation, and industry?*

Report	Response*	Page	Comments
MOI.3B	Yes (Roughly)	p. 8	For United States in 1980: Electric utilities 29 percent, Industrial 18 percent, Transportation 44 percent.
	No	p. 10	For Canada in 1980: Transportation 60 percent, Industrial 16 percent, Electric utilities 14 percent.
NAPAP	Yes** (Roughly)	p. 20	"Estimated man-made emissions of NO$_x$ in 1977 ... [were] 39 percent ... from transportation, and ... 61 percent from combustion in stationary sources."
CAD.1	No	p. 2.52	"Within the last decade, mobile sources and electric utilities have been the predominant contributors."
DYCO	No***	p. 7	"Transportation and stationary source fuel combustion (e.g., power plants) are the principal sources of the oxides of nitrogen." (For 1977, transportation produced 40 percent, stationary sources 56 percent, and industrial processes only 3 percent.)
NRCC	Yes (Roughly)	p. 58	"Approximately 27% of North American NO$_x$ emissions is attributed to power plants, whereas 31 percent is attributed to industrial, commercial and residential fuel combustion. Approximately 41 percent of North American NO$_x$ emissions (is) due to transportation sources."
NRC.1	Yes (Roughly)	p. 41	"National NO$_x$ emissions ... [in the United States] arise in roughly equal proportions from automobiles, industry and electric utilities."
NRC.2	Not addressed		
ON-HY	Yes	p. 8	For United States: "In 1977, transportation was still the major source at more than 40 percent of the total, with electrical utilities at about 30 percent and other stationary sources near 25 percent."

Table A-5.   (continued)

Report	Response*	Page	Comments
	No	p. 8	For Canada: "The transportation sector contributed a higher proportion of the total emissions in Canada and in Ontario, about 60 percent, while electric utilities were small emitters at about 10 percent."
SCAR	Yes (Roughly)	p. 19	United States and Canadian figures taken together, emissions accounted for 27 percent, industrial boilers/process heaters/residential/commercial accounted for 31 percent, and transportation for 41 percent.

*Yes = Equally electric utilities, transportation, and industry; No = Other distribution of sources.

**Yes, assuming "stationary sources" are taken to include equal amounts of emissions from power plants and industry.

***No, assuming "industrial processes" are defined as nonferrous metal smelters, iron and steel mills, petroleum refineries, cement quarries, pulp and paper plants, asphalt and chemical works, and manufacturers of soap and synthetic detergents, glass fiber, and textiles (page 13 of same document).

Table A-6.   Natural and Anthropogenic Emissions

Question 3.a. *Have natural sources of nitrogen and sulfur compounds,
though dominant sources globally, become less significant when compared
with anthropogenic emissions on a regional scale in North America?*

Report	Response	Page	Comments
MOI.H	Yes	p. 2	"In the region of concern, natural sources of sulphur cannot account for sulphur levels observed in surface waters. It is estimated that man-made sulphur dioxide emissions exceed natural emissions by a factor of 10 to 20."
NAPAP	Yes	p. 20	"Globally, the relative magnitudes of natural and man-made emissions of the two gases are uncertain. In the eastern U.S. and adjacent parts of Canada, however, man-made emissions of sulfur dioxide are estimated to be substantially larger than natural emissions."
CAD.1	Yes	p. 38	"For sulfur compounds, the contributions of natural sources appear to be insignificant when compared to anthropogenic emissions."
		p. 75	"Perhaps about 10% of the NO_x contribution to acid precipitation may be due to natural NO_x sources."
DYCO	Yes	p. 5	"Taken on a worldwide basis, natural emissions of some pollutants . . . often exceed anthropogenic . . . emissions by several orders of magnitude. The reason man-made emissions so noticeably affect the quality of the environment is that they tend to concentrate locally or regionally in air masses downwind of urban areas or large point sources."
NRCC	Yes	p. 51	"Anthropogenic emissions . . . are thought to account for slightly more than 50% of the total annual global emissions of sulfur to the atmosphere, and . . . for an uncertain but smaller percentage of the total nitrogen oxides emissions. . . . [However], in . . . eastern North America, anthropogenic emissions account for more than 90% of the total (anthropogenic and

Table A-6.   (continued)

Report	Response	Page	Comments
			natural) sulfur emissions, and a relatively large but unknown percentage of the total nitrogen oxides emissions."
NRC.1	Yes	p. 147	"Emissions of nitrogen and sulfur oxides and the consequent acid precipitation are broad regional rather than global problems. When natural and anthropogenic emissions are compared on a regional basis, it is clear that man's activities completely overwhelm natural sources of SO_2 and NO_x, even though the magnitude of anthropogenic emissions of these oxides may seem unimportant when compared with natural emissions on a global scale."
NRC.2	Yes	p. 13	"More than half the acidity of precipitation averaged over the globe may be due to natural sources, but anthropogenic sources may dominate in some regions. For example, in eastern North America 90 to 95 percent of precipitation acidity may be the result of human activities, although natural sources may also be important at times in specific locations."
ON-HY	Yes	p. 9	"Globally, natural sources of nitrogen and sulfur compounds are considered to be dominant but they become less significant when compared to anthropogenic emissions on a regional scale such as the highly industrialized northeastern United States."
SCAR	Not addressed	p. 19	Anthropogenic emissions are not discussed in the context of natural emissions.

Table A-7.   Dry and Wet Deposition

Question 3.b. *Is the damage resulting from dry acid deposition comparable to that of wet?*

Report	Response	Page	Comments
MOI.2	Yes	p. 11.4	"Studies . . . suggest that in eastern North America the three removal terms, wet deposition, dry deposition, and outflow (into the Atlantic) are roughly equal. Further work is underway to determine the validity of this estimated apportionment."
NAPAP	Not addressed		
CAD.1	Not addressed		
DYCO	Don't know	p. 34	"Rates of dry deposition are virtually impossible to predict with any degree of assurance since they are controlled by numerous factors."
NRCC	Yes	p. 34	"Although problems inherent in dry deposition measurement render precise knowledge of dry deposition rates difficult, both wet and dry deposition are thought to be of approximately equal importance on a regional basis in Canada."
NRC.1	Not addressed		
NRC.2	Yes	p. 46	"Roughly one third of northeastern emissions is assumed to be dry-deposited on the North American continent. . . . Difficulties combine to give a number of widely varying estimates for the temporal and spatial scales of dry deposition of specific pollutants. As a rule of thumb, for sulfur and nitrogen compounds at least, dry deposition is taken on the average to be about as effective as wet deposition in pollutant removal. About one third of sulfur emissions is transported out of the continent. Thus roughly one third of northeastern emissions is assumed to be dry-deposited on the North American continent."

Table A-7.   (continued)

Report	Response	Page	Comments
ON-HY	Yes	p. 75	"Dry deposition of these compounds may be of equal or greater significance than wet deposition in many areas."
SCAR	Almost	p. 68	"Some scientists estimate that 10 to 30 percent of the acid precipitation problem may result from dry deposition. . . . We believe that this component of acid precipitation is sufficiently important to justify a major and continuing research effort."

Table A-8.   Long-Distance Transport

Question 4. *Can the precursors of acid deposition, or the resulting acids, be transported long distances from their sources (including to previously unaffected areas)?*

Report	Response	Page	Comments
MOI.2	Yes	p. 11.14	"Deposition values at the more remote pristine locations in eastern Canada and in other remote areas worldwide clearly cannot be attributed to local sources, which are negligible, and demonstrate the reality of a long range component."
NAPAP	Yes	p. 25	"Depending upon meteorological conditions and circumstances of release, sulfur and nitrogen oxides can either be dispersed locally or transported hundreds or perhaps thousands of miles from their point of origin."
CAD.1	Yes	p. 4.1	"The regional distribution of secondary pollutants such as sulfates in . . . North America has been recognized to be a consequence of long-range transport and chemical transformations of pollutant emissions into the atmosphere."
DYCO	Yes	p. 29	"Substances are subject to short-range, intermediate-range, and long-range transport."
NRCC	Yes	p. 50	"The phenomenon is significant on the scale of hundreds to thousands of square kilometres, in large part because the pollutant materials (primarily sulfur and nitrogen oxides) which are emitted into the atmosphere in massive quantities are slowly removed during the time required for the regional winds to transport them across the extent of such regions."
NRC.1	Yes	p. 77	"It is clear that a large number of toxic substances are emitted to the atmosphere and transported long distances, only to be redeposited in quantities large enough to justify some concern."
NRC.2	Yes	p. 140	"Both observational and theoretical evidence exists for the long-range transport of pollutants leading to acid deposition. . . .

Table A-8.   (continued)

Report	Response	Page	Comments
			Any receptor site will be influenced to one degree or another by both local and distant sources."
ON-HY	Yes	p. 11	"Obviously, particles and gases emitted into the atmosphere can traverse long distances."
SCAR	Yes	p. 13	"The meteorological patterns in eastern North America dictate that atmospheric pollutants can move over great distances within Canada and the United States, and also across the international border.'

Table A-9.   Tracing Point Sources

Question 5. *Can damage to specific areas be attributed to point sources of emissions?*

Report	Response	Page	Comments
MOI.2	No	p. 11.3	"While . . . analyses reinforce the conclusion that man-made emissions have a major influence on acidic deposition, this method is unable to distinguish between near and more distant sources within the same directional sector and cannot be used to trace an air mass trajectory during periods of weak, variable air flows or over very long distances."
NAPAP	No	p. 25	"Although such long-range transport is becoming better understood, it is still not possible to determine the extent to which any specific source or collection of sources in one region leads to acid deposition in another region."
CAD.1	No	p. 4.1	"Within the range of the mesoscale a specific plume from a power plant or urban complex will commonly lose its identity by mixing with other plumes or by diluting indistinguishably into the background."
DYCO	No	p. ix	Discussion that current predictive models are not adequate to trace acid precipitation from sources to areas of deposition.
NRCC	Not addressed		
NRC.1	No	p. 77	"[Before relationships can be predicted] . . . we need . . . a greatly improved understanding of the mechanisms by which materials are emitted to the atmosphere, transported from point sources over long distances by local winds and air-mass ⌐movements, and deposited in specific ecosystems."
NRC.2	No	p. 9	"The relative importance for deposition at specific sites of long-range transport from distance sources as compared with more direct influence of local sources cannot be determined from currently available data

Table A-9.   (continued)

Report	Response	Page	Comments
			or reliably estimated using currently avail-able models."
ON-HY	No	p. 19	"The linking mechanisms between sources and receptors of pollutants several hundred kilometres apart are not well understood."
SCAR	No	p. 13	"A major difficulty in effecting control of emissions from a particular industrial site is the lack of conclusive evidence linking acid rain in any specific area with a specific source."

Table A-10.   Linear Relationship between Emissions and Deposition

Question 6. *Is the relationship between emissions and acid deposition linear?*

Report	Response	Page	Comments
MOI.2	No	p. 11.15	"Over the shorter time and space scales, all of the important sulfur dioxide chemical conversion processes are non-linear. Current long range transport model studies by [this work group] make linear approximation of these various chemical processes and may be subject to error in the prediction of the depositions of individual sulfur species."
NAPAP	Don't know	p. 29	"The relationship between the emissions of acid could be nonlinear, so that a considerable reduction in emissions might not result in a similar reduction in acid precipitation."
CAD.1	Not addressed		Though both modeling and transformation are discussed, linearity between emissions and acid deposition specifically is not.
DYCO	Don't know	p. 32	"Known reactions alone are insufficient to explain SO_2 conversion to SO_4 in the patterns and quantities observed."
NRCC	Not addressed		Though both modeling and transformation are discussed, linearity between emissions and acid deposition specifically is not.
NRC.1	Not addressed		
NRC.2	Yes	p. 7	"There is no evidence for a strong non-linearity in the relationships between long-term average emissions and deposition [in North America]."
ON-HY	Not addressed		Though both modeling and transformation are discussed, linearity between emissions and acid deposition specifically is not.
SCAR	Not addressed		

Table A-11.   Accuracy of Modeling

Question 7. *Are either transformation or transportation models accurate enough to be relied upon for development of control strategies?*

Report	Response	Page	Comments
MOI.2	Maybe	pp. 11.6–7	"Of eight long range transport models . . . different models exhibit variations. . . . This variability could lead to substantial differences in the selection of optimum emissions reduction scenarios depending on the particular model applied and the level of detail required. . . . It has not been possible to date to choose a 'best model' among the eight."
NAPAP	No	p. 36	"To construct models that can be useful to decision-making concerning acid precipitation, more must be known."
CAD.1	No	p. 4.74	"Transformation models can, at best, be only as good as our understanding of the transformation processs. Significant gaps in this understanding remain."
DYCO	No	p. viii	"The simulation model [devised for this report] . . . was not intended to be a predictive management tool and, at its current level of refinement, use of the model for predictive purposes is inappropriate."
		p. ix	"Models . . . [are] incomplete representations of real world systems and, as such, should be only one of the tools used to make management decisions."
NRCC	Not addressed		
NRC.1	Not addressed		
NRC.2	No	p. 10	"For choosing among possible emission-control strategies . . . we do not believe it is practical at this time to rely upon currently available models . . . In the absence of other methods, analysis of observational data provides guidance for assessing the consequences for changing SO_2 emissions for wet deposition of sulfate."

Table A-11.   (continued)

Report	Response	Page	Comments
ON-HY	Yes	p.78	"Despite incomplete and inadequate knowledge of the processes involving the precursors of acid deposition, numerical models can be constructed which provide sufficient information to guide consideration of options in designing a control strategy to minimize the effect of acid deposition. Model data can also guide the selection of control options to maximize the environmental benefits derived from emission control programs."
SCAR	Not addressed		

Table A-12.   Future Increases in Deposition

Question 8. *Are increases in acid deposition or its precursors predicted for the future?*

Report	Response	Page	Comments
MOI.3B	Yes	p.10	SO_2 will increase 10 percent for United States and decrease 5 percent for Canada from 1980 to 2000. NO_x will increase 25 percent for United States and 32 percent for Canada from 1980 to 2000.
NAPAP	Yes	p. 20	"It is estimated that over the next 20 years, annual emissions of man-made SO_2 will increase slightly to about 30 million tons, with approximately 18 million tons expected to come from electric utilities. NO_x emissions are anticipated to increase roughly 16 percent, to about 27 million tons per year. This increase is . . . expected to come . . . from stationary sources [rather than transportation]."
CAD.1	Not addressed		
DYCO	Yes	pp. 8, 10	Though anthropogenic emissions of hydrocarbons will remain relatively constant from 1985 to 2000, NO_x emissions will rise approximately 25 percent during that period. (As interpreted from provided histograms.)
NRCC	Yes	p. 58	Yes, for North America as a whole, due to Canada: "In the next two decades, United States emissions from power plants are predicted to remain relatively constant. In Canada, SO_2 emissions from power plants are expected to increase 75% by 2000."
			"Nitrogen oxides emissions from the transportation sector in the United States are not expected to vary significantly in the next 15–20 yr. In the absence of further controls, NO_x emissions from transportation in Canada are expected to increase by approximately 60%."
NRC.1	Yes	p. 47	"Significant increases in NO_x emissions are forecast for the remainder of the

Table A-12.   (continued)

Report	Response	Page	Comments
			century, primarily due to emissions associated with increased coal use in the electric utility and industrial sectors of the U.S. economy."
		p. 41	"Even with more stringent standards applied to new power plants SO_2 emissions are predicted to continue at about their present level."
NRC.2	Not addressed		
ON-HY	Not addressed		
SCAR	Yes	p. 14	"Evidence suggests that continued pollutant loading of sensitive areas at current levels will result in progressive deterioration of the environment. Moreover, the effects will become extensive and irreversible over the next 10 to 20 years, particularly in aquatic ecosystems."

Appendix B
Detailed Assessment Responses:
Environmental Effects

Table B-1.   Damage to Aquatic Ecosystems

Question 1. *Does acidification substantially damage aquatic ecosystems?*

Report	Response	Page	Comments
OTA	Yes	pp. 44-47	
MOI.1c	Yes	p. 1.4	"Observed changes in aquatic life have been correlated with measured changes in the pH of water and compared for waters of different pH values. Differences have been documented in species composition and dominance and size of plankton communities in lakes of varying pH. . . . These alterations may have important implications for organisms higher in the food chain."
MOI.1us	Perhaps	p. 1.10	"Observed changes in aquatic life have both been correlated with measured changes in the pH of water and inferred by comparisons of waters of different pH values. Differences have been documented in species composition and dominance and size of plankton communities in lakes of varying pH. . . . These differences may have important implications for organisms higher in the food chain, but studies to date have not been done that might establish this connection."
NAPAP	Yes	p. 37	"It is known that many lakes . . . are showing stress from increased acidity, and that

Table B-1.   (continued)

Report	Response	Page	Comments
			lakes and streams elsewhere . . . also appear to be vulnerable."
ARIB	Yes	p. 177	"Data . . . suggests that acid precipitation, by acidification of lakes, has been responsible for the elimination of acid-sensitive aquatic species and has disrupted primary production and the nutritional food web within the affected ecosystem."
CAD.2	Yes	Chap. E-5 p. 215	"Acidification produces changes in the basic structure of aquatic ecosystems."
DYCO	Yes	p. 106	"The effects of acid deposition in aquatic ecosystems are numerous, complex, varied, and highly interrelated. In regions sensitive to acid inputs, virtually every respect of ecosystem structure and function is subject to alteration."
NRCC	Yes	p. 29	"Acidification and associated chemical, physical, and biological changes are evident in aquatic ecosystems in North America."
		p. 32	"The situation facing the aquatic environment in Canada today appears critical."
NRC.1	Yes	p. 161	"As a result of acidification a number of organisms have been reduced or eliminated over significant parts of their ranges."
NCSU	Yes	p. 5	"As of 1980, hundreds of lakes . . . [in U.S. and Canada] . . . were showing acid stress in the form of diminished populations or extinction of fish populations."
EPRI	Maybe	p. 4	Experiments suggest it can.
EEI	Maybe	p. 29	"There is no question that the acidity of precipitation can affect the lakes into which it falls or flows, but . . . the varied and complex interacting influences make it impossible to pinpoint any single factor as being the major determinant of lake acidification or fish loss."
LAW	Yes	p. 27	"The most dramatic . . . impacts have been visited upon lakes and streams in areas

Table B-1. (continued)

Report	Response	Page	Comments
			downwind of major industrial centers and low in natural buffers."

Note: In Appendix B tables italics have
been added for emphasis.

Table B-2.   Vegetation Damage

Question 2. *Does acid deposition substantially damage terrestrial vegetation?*

Report	Response	Page	Comments
OTA	Don't know	p. 1.14	To date, there have been too few studies to establish a clear relationship on interactions of acidic deposition/sulfur dioxide/ozone to reach a definitive conclusion on effects.
MOI.1	Likely	p. 1.14	Few direct effects have been documented in the field. But because studies have demonstrated direct effects on soils, and because forested areas are generally associated with areas of lower pH soils with low nutrients status, any further loss of nutrients is considered significant, however small that may be.
NAPAP	Don't know	p. 39	"Laboratory studies suggest that excess acid deposition can have a number of adverse effects [on forest ecosystems]."
ARIB	Don't know	p. 177	"To date, there has been no visible or detectable damage to terrestrial ecosystems outside the laboratory."
CAD.2	Don't know	Chap. E-3 p. 61	"Although at present we have no conclusive proof that acidic deposition currently limits forest growth in either the United States or Europe, we do have indications that growth reductions are occurring, principally in coniferous species that have been examined to date, that these reductions are rather widespread, and that they occur in regions where rainfall acidity is generally quite high."
DYCO	Probably	p. 98	"The dominant ecosystem-level effects of the acidifying air pollutants and acid precipitation involve alterations of soil chemistry, fertility and structure, which can diminish the productivity of the habitat."
NRCC	Not addressed		
NRC.1	Don't know	p. 167	"Adverse effects of acid precipitation have not been proven."

Table B-2.   (continued)

Report	Response	Page	Comments
NCSU	Yes	pp. 5, 6	"In certain industrial regions of the world, substantial damage to forests and agricultural crops is caused by the dry deposition of toxic gases." Also, laboratory experiments indicate it can.
EPRI	Don't know	pp. 2-4	
EEI	Don't know	p. 28	"Fears that acid rain can damage forestry resources have been voiced but not substantiated."
LAW	Likely	pp. 30, 31	"There is evidence to suggest that forests in widespread areas of . . . North America may be suffering serious, perhaps fatal, acid-induced damage. However, documentation is lacking, and we are only beginning to understand the scope and nature of possible impacts."

Table B-3.   Soil Damage

Question 3. *Does acid deposition substantially damage soils?*

Report	Response	Page	Comments
OTA	Yes	Sec. F, p. 8	However, damage to soils is often in locations that will little affect forest ecosystems; yet damage will result to aquatic ecosystems.
MOI.1	Don't know	p. 1.16	"A variety of soil effects have been demonstrated, usually of an undesirable nature, but at the present time, the problem remains of quantifying the dose-response reactions in . . . field situations."
NAPAP	Not addressed		
ARIB	Don't know	p. 173	
CAD.2	Likely	Chap. E-2 p. 49	"Increased mobility of Al in uncultivated, acid soils is probably the most significant effect of acid deposition on soils as they influence terrestrial plant growth and aquatic systems."
DYCO	Yes		"The dominant ecosystem-level effects of the acidifying air pollutants and acid precipitation involve alterations of soil chemistry, fertility and structure."
NRCC	Not addressed		
NRC.1	Yes	p. 170	"Acid rain has particularly severe effects upon the cycles of metallic toxicants and nutrients."
NCSU	Yes	p. 6	"Soil acidification increases leaching of exchangeable plant nutrients . . . increases rate of weathering of most minerals . . . decreases the rate of many soil microbiological processes such as N fixation and breakdown of organic matter . . . inhibits certain microbial processes in forest litter, especially decomposition of organic matter . . . [and] reduces the availability of phosphorus to plants and increases the solubility of other elements, some of which may be toxic to plants."

Table B-3.   (continued)

Report	Response	Page	Comments
EPRI	Don't know	pp. 2–5	
EEI	Not addressed		
LAW	Probably	p. 32	"Acid-induced changes in the forest soil can lead to important forestry impacts."

Table B-4.   Major Damage Factors in Aquatic Ecosystems

Question 4. *What are the main causes of damage to the aquatic ecosystem with regard to acid deposition?*

Report	Page	Response
OTA	Sec. H, p. 1	The synergistic effects of: a) acidity b) presence of metals c) type of acid input
MOI.1	pp. 1.2, 4	a) pH depressions b) nitrate concentrations c) metal accumulations in surface waters
NAPAP	p. 38	a) acidity b) toxic metal ions, particularly aluminum
ARIB	p. 164	a) acidity b) presence of such components as heavy metals (aluminum, manganese, mercury), and carbon dioxide
CAD.2	Chap. E-5, pp. 209, 214, 215	a) acidity b) elevated metal concentrations c) reduced phosphorus levels
DYCO	pp. 116, 117	a) acidity b) metal concentrations
NRCC	a) p. 256 b) pp. 264, 265	a) acidity b) possibly aluminum leached from surrounding soils
NRC.1	a) p. 149 b) p. 157	a) acidity b) aluminum concentrations due to leaching
NCSU	p. 5	a) acidity b) certain cations, particularly aluminum
EPRI	pp. 2–6	a) probably not due to acid deposition per se; rather, perhaps damage is a product of overall water quality b) perhaps metal toxicity
EEI	p. 31	Lake chemistry is influenced strongly by too many other factors to be able to isolate the interactions of acid deposition.
LAW	p. 29	Toxic combinations of water acidity and aluminum ions

Table B-5.   Natural Reversal of Damages

Question 5. *Given a decrease or termination of acid deposition, would the incurred adverse effects reverse naturally with time?*

Report	Response	Page	Comments
OTA	Partly	Sec. G, p. 13	
MOI.1C	Partly	p. 1.7	"Reductions from present levels of total sulfur deposition would reduce damage to sensitive (low alkalinity) surface waters and would lead to eventual recovery of those waters that have already been altered chemically or biologically. *Loss of genetic stock would not be reversible.*"
MOI.1US	Partly	p. 1.12	"Reductions from present levels of total sulfur deposition would reduce further chemical and biological alterations to low alkalinity surface waters currently experiencing effects and would lead to eventual recovery of those waters that have been altered by deposition."
Joint	Maybe	p. 3.144	"Loss of fish populations with specific gene characteristics from lakes and rivers may be an irreversible process."
		p. 3.146	"Evidence seems to be conflicting as to whether the geochemical alteration of watersheds due to acidic input should be viewed as irreversible, and, if so, on what scale."
		p. 3.146	"The loss of soil cations . . . is another potential irreversible consequence. . . . However, the extent to which [this is] . . . significant . . . is unknown."
NAPAP	Not addressed		
ARIB	Not addressed		
CAD.2	Not addressed		
DYCO	Not addressed		
NRCC	Possibly	p. 29	"Ecosystem effects are . . . possibly irreversible."

Table B-5.   (continued)

Report	Response	Page	Comments
NRC	Not addressed		
NCSU	Not addressed		
EPRI	Not addressed		
EEI	Not addressed		
LAW	Probably not	p. 30	"No natural means is known by which an acidified lake might return to its original chemical and biological composition."

Table B-6.   Liming

Question 6. *For acidified aquatic ecosystems, does liming appear to be a feasible remedy?*

Report	Response	Page	Comments
OTA	Not addressed		
MOI.1	Sometimes	pp. 1.23—24	"Liming will not eliminate all problems associated with acidification of surface waters but may be necessary on a limited basis as a means of temporarily mitigating the loss of important aquatic ecosystem components. However, it cannot be used in all situations. Further, its long-term viability and impact on fish populations needs additional study."
NAPAP	Not addressed		
ARIB	Possibly	p. 192	"Further studies are required to assess the true feasibility of this approach and the methodologies that would need to be applied. While increases in total biomass have been observed as a result of liming, other studies indicate liming compounds the effects of aluminum toxicity and cadmium accumulation. Costs may also be prohibitive."
CAD.2	Somewhat	Chap. E-5 p. 204	"Short-term effects of neutralization includes aluminum hydrolysis, which is detrimental to existing fish populations." "The long-term consequence of lake neutralization, provided reacidification is not allowed to occur, is a much more hospitable environment for fish."
		Chap. E-5 p. 202	Initially, both phytoplankton and zooplankton densities decline drastically following neutralization. The former recovers almost fully within a few months; zooplankton recovery takes two full years, however. Neutralization also has strong effects on benthic fauna.
DYCO	Not addressed		

Table B-6.   (continued)

Report	Response	Page	Comments
NRCC	Possibly	pp. 47, 265	"Some effects of acidification may in certain cases be partially halted or reversed. However, the extent of recovery of the biotic community is apparently ultimately limited by the loss of species during the acidification process." Further, "potentially toxic aluminum may be liberated upon liming, thus inducing fish kills."
NRC.1	Don't know	p. 182	"pH and alkalinity cannot be artificially maintained without increasing the ionic concentration of the receiving water, the consequences of which have not been investigated. The fate of dissolved toxic metals after liming is also poorly known."
NCSU	Not addressed		
EPRI	No	p. 12	
EEI	Yes	p. 39	"Liming efforts have been increasingly promising."
LAW	No	p. 78	May interfere with natural processes of nitrogen circulation in forest soils, and may liberate toxic metals in the water.

Table B-7.   Acid Shock versus Gradual Impacts

Question 7. Does the impact of acid shock as a product of snowmelts result in more damage to aquatic chemistry and/or ecosystems than if the same impact had been gradual?

Report	Response	Page	Comments
OTA	Yes	Sec. H, p. 1	"Rapid decreases in stream and lake pH due to spring snowmelt and release of acid accumulated over the winter can be detrimental if they coincide with sensitive periods of the fish reproductive cycle."
MOI.1	Yes	p. 3.92	"If short-term changes in water chemistry coincide with sensitive periods in the life cycles of fish, *significant* mortality and reduced reproduction can occur."
NAPAP	Yes	p. 37	"Sudden increase or 'surges' in acidity are generally more harmful to fish than gradual changes.... Short-term releases of pollutants (due to snowmelt surges) can cause major and rapid changes in the acidity and other chemical properties of the receiving waters. Fish kills have been a dramatic consequence of such episodes."
ARIB	Maybe		
CAD.2	Yes	Chap. E-5 p. 209	"Episodic depressions down to pH 3.8 to 4.6 ... during periods of snowmelt and heavy rainfall ... will have *significant* harmful effects on aquatic organisms."
DYCO	Yes	p. 50	"Snowmelts are an important hydrological factor in ecosystem vulnerability as they may cause rapid release of the pollutants and acids that have accumulated over the course of a season."
NRCC	Yes	p. 43	"Short-term changes (due to snowmelt surges) ... may have a *significant* effect upon aquatic biota.... Short-term pH depressions, and in some cases elevated metal levels in surface waters, are probably widespread phenomena."
NRC	Yes	p. 99	"It is now recognized that fish mortality attributed to acid precipitation is *in large part* caused by the aluminum leached from terrestrial soils during spring melt."

Table B-7.   (continued)

Report	Response	Page	Comments
NCSU	Yes	p. 4	"The . . . release of pollutants [during snow-melts] can cause major and rapid changes in the acidity and other chemical properties of stream and lake waters. *Fish kills* are a *dramatic* consequence of such episodic inputs into aquatic ecosystems."
EPRI	Not addressed		
EEI	Yes	p. 29	"A sudden flush of activity may *overwhelm* the natural buffering capabilities of a body of water and significantly lower its pH for a period of time."
LAW	Yes	p. 29	"*Dramatic large-scale fish kills* may result from 'acid shock' which accompanies the thawing of acid-laden snow."

Table B-8.   Highest Admissible Acid Loading

Report	Page	Response
EEI	p. 38	"There is no evidence that reduction in *emissions* would have significant effect on the acidity of precipitation."
NRC		"It is desirable to have precipitation with pH values no lower than 4.6 to 4.7 throughout [sensitive aquatic] areas."
MOI.1US	p. 1.12	"It is not now possible to derive quantitative loading/effects relationships."
MOI.1C	p. 1.7	"Deposition of sulphate precipitation (could/ should) be reduced to less than 20 kg/ha/yr in order to protect all but the most sensitive aquatic ecosystems in Canada."

Table B-9.   Future Trends of Acid Deposition

Report	Page	Response
OTA	Sec. G, pp. 12, 13	a) If sulphate deposition was increased 10 percent by the year 2000, conditions in 5 to 15 percent of the most sensitive lakes would worsen.
		b) If decreased by 20 percent, 10 to 25 percent of said lakes would experience some recovery.
		c) If decreased by 35 percent, 14 to 40 percent of said lakes would experience some recovery.
NRC.1	pp. 2, 3	Without action there will be increased damage; perhaps acid deposition problems should be seen as part of a larger problem—the use of fossil fuels, resulting in environmental consequences that might prove very severe.

Notes

Chapter One

1   Numerous summaries of scientific information on acid deposition are available, among them the scientific-assessment documents analyzed in chapter 2. The following recent publications provide useful overviews of the main findings to date: Michael Oppenheimer, *Reducing Acid Rain: The Scientific Basis for an Acid Rain Control Policy* (New York: Environmental Defense Fund, 1984); Frank A. Record, David V. Bubenick, and Robert J. Kindya, *Acid Rain Information Book* (Park Ridge, N.J.: Noyes Data Corporation, 1982); U.S. Congress, Office of Technology Assessment (OTA), *Acid Rain and Transported Air Pollutants: Implications for Public Policy* (Washington, D.C.: Government Printing Office, 1984); James White, ed., *Acid Rain: The Relationship between Sources and Receptors* (New York: Elsevier Science Publishing, 1988); Ernest Yanarella and Randal Ihara, eds., *The Acid Rain Debate: Scientific, Economic, and Political Dimensions* (Boulder, Colo.: Westview Press, 1985).

2   National Academy of Sciences, National Research Council, Committee on Atmospheric Transport and Chemical Transformation in Acid Precipitation, *Acid Deposition: Atmospheric Processes in Eastern North America* (Washington, D.C.: National Academy Press, 1983). Hereinafter National Academy of Sciences will be abbreviated NAS.

3   Oppenheimer, *Reducing Acid Rain*, p. 9.

4   Jeremy Hales, "Atmospheric Chemistry—A Lay Person's Introduction," in *Acid Rain: The Relationship between Sources and Receptors*, ed. James White (New York: Elsevier Science Publishing, 1988), pp. 71–89.

5   The following discussion of different stages in environmental policy is based on developments in the United States only.

6   This is the number of entries in EPA's Toxic Substances Control Act registry.

7   U.S. Office of Science and Technology Policy, Executive Office of the Presi-

dent, *General Comments on Acid Rain, A Summary by the Acid Rain Peer Review Panel for the Office of Science and Technology Policy*, June 27, 1983, p. 4.

8   These reports were prepared jointly by Canadian and U.S. experts under a memorandum of intent signed in 1978. For details see chapter 3.

9   See Jurgen Schmandt, "Regulation and Science," *Science, Technology, and Human Values* 9, no. 1 (Winter 1984): 23–38.

10   See chapter 2.

11   U.S. Congress, Budget Office, *Curbing Acid Rain: Cost, Budget, and Coal-Market Effects* (Washington, D.C.: Government Printing Office, 1986), pp. xix–xxvii.

12   OTA, *Acid Rain and Transported Air Pollutants*, p. 3.

13   Larry Parker, "Summary and Analysis of Technical Hearings on Costs of Acid Rain Bill," Congressional Research Service, Library of Congress, July 26, 1982, mimeo, pp. 28–29. This paper is an excellent summary of economic studies on acid rain.

14   See chapter 7.

15   More stringent automobile emission standards in California than in the rest of the nation have been the major exception to this rule.

16   OTA, *Acid Rain and Transported Air Pollutants*, p. 169.

17   See chapter 8.

18   See the discussion on the special envoys' report in chapter 3.

19   Irreversibility of effects is a reasonable hypothesis, which, however, has not yet been documented.

20   Amendments to the U.S. Clean Air Act prevent continued use of this evasive strategy. Canada should consider taking similar steps.

Chapter Two

1   Joan P. Baker, "Acidic Deposition and its Effects on Aquatic Ecosystems," unpublished report, Duke University, AAAS/EPA Environmental Science and Engineering Program, 1982.

2   National Academy of Sciences, Committee on Monitoring and Assessment of Trends in Acid Deposition, *Acid Deposition: Long-term Trends* (Washington, D.C.: National Academy Press, 1986).

3   National Acid Precipitation Assessment Program (NAPAP), *Interim Assessment: The Causes and Effects of Acid Deposition*, 1–4 (Washington, D.C.: U.S. Department of Commerce, 1987).

4   Ad Hoc Committee on Acid Rain "Is There Scientific Consensus on Acid Rain?, Excerpts from Six Government Reports," October 1985.

5   Most industry-sponsored assessments known to the author have been produced by private utility companies. Yet industry reports were often difficult to obtain, and this review, therefore, may not fully represent the positions of private industry. Several assessments would have been useful but could not be obtained, in particular P. F. Chester, *Effects of Sulfur Dioxide and its Derivatives on Health and Ecology* (Leatherhead,

England: Central Electricity Generating Board, 1981) and W. C. Retzsch, A. G. Everette, P. F. Dunhaime, and R. Nothwanger, *Alternative Explanations for Aquatic Ecosystems Effects Attributed to Acidic Deposition* (Rockville, Md.: Utility Air Regulatory Group, 1982).

6   Only Canadian government documents (as opposed to reports prepared by Canadian industry or independent research groups) were available for the section on environmental effects.

7   National Research Council, Committee on Atmospheric Transport and Chemical Transformation in Acid Precipitation, *Acid Deposition: Atmospheric Processes in Eastern North America* (Washington, D.C.: National Academy Press, 1981), pp. 45–46.

8   National Research Council, Committee on the Atmosphere and the Biosphere, *Atmosphere-Biosphere Interactions* (Washington, D.C.: National Academy Press, 1981), p. 35.

9   Residence time refers to a given compound's propensity to remain aloft relative to other compounds.

10 United States–Canada, *Memorandum of Intent on Transboundary Air Pollution: Atmospheric Sciences and Analysis* (Final draft: November 1982), p. 11.5.

11 Gregory Wetstone and Armin Rosencranz, *Acid Rain in Europe and North America: National Responses to an International Problem* (Washington, D.C.: Environmental Law Institute, 1983), p. 12.

12 Ibid., p. 21.

13 Canada, House of Commons, Subcommittee on Acid Rain, *Still Waters: The Chilling Reality of Acid Rain* (Ottawa: Minister of Supplies and Services, 1981), p. 13.

14 G. Tyler Miller, Jr., *Living in the Environment* (Belmont, Calif.: Wadsworth Publishing, 1979), p. A9.

15 Baker, "Acidic Deposition and its Effects on Aquatic Ecosystems."

16 Robert M. Friedman, et al. *The Regional Implications of Transported Pollutants: An Assessment of Acidic Deposition and Ozone* (Washington, D.C.: Office of Technology Assessment, 1982), p. F-12.

17 Wetstone and Rosencranz, *Acid Rain in Europe and North America*, p. 32.

18 Ibid., pp. 32–33.

19 Frank A. Record, David V. Bubenick, and Robert J. Kindya, *Acid Rain Information Book* (Bedford, Mass.: GCA Corporation, 1982), p. 164.

20 Baker, "Acidic Deposition and Its Effects on Aquatic Ecosystems," p. 1.

21 In addition to the two mentioned in the analysis, they include NAS, *Acid Deposition: Processes of Lake Acidification* (Washington, D.C.: National Academy Press, 1984), and NAS, *Acid Deposition: Long-Term Trends*.

22 NAS, *Acid Deposition,* pp. 6-9.

23 Ad Hoc Committee, "Is There Scientific Consensus on Acid Rain?"

24 NAPAP, *Interim Assessment.*

25 Federal/Provincial Research and Monitoring Coordinating Council, "A Critique of the U.S. National Acid Rain Assessment Program's Interim

Assessment Report," December 1987.
26   Ad Hoc Committee, "Is There Scientific Consensus on Acid Rain?"

Chapter Three

1   G. V. LaForest, "Boundary Waters Problems in the East," in *Canada–United States Treaty Relations*, ed. David R. Deener (Durham, N.C.: Duke University Press, 1963), pp. 34–49.
2   Foreign Relations Re-authorization Act of 1978, P.L. 95-426, 92 Stat. 963, Section 612, October 7, 1978.
3   Stephen Clarkson, *Canada and the Reagan Challenge* (Toronto: James Lorimer, 1982), p. 187.
4   This group and its work will be discussed in greater detail in chapter 4.
5   "Canada: Transboundary Air Quality Talks," *Department of State Bulletin* (January 1980): 4.
6   Boundary Waters Treaty of 1909, Department of State Treaty Series 548, Article 5.
7   "Canada: Transboundary Air Quality Talks," *Department of State Bulletin* (November 1979): 26–27.
8   John E. Carroll, *Acid Rain: An Issue in Canadian-American Relations* (Ontario: C.D. Howe Institute, 1982), pp. 39–40.
9   U.S. Department of State, *Transboundary Air Pollution: Memorandum of Intent between the United States of America and Canada, Treaties and Other International Acts*, series 9856 (Washington, D.C.: Government Printing Office, 1980), p. 4.
10 Ibid., p. 56.
11 "Relations with the United States: Acid Rain," *International Canada* (March 1981): 78–79.
12 A study by the Congressional Research Service concluded that the charge was without foundation because technical, regulatory, and political reasons make it virtually impossible to substitute Canadian electricity in a significant way for U.S. coal-fired capacity. See Larry Parker, "Acid Rain Legislation and Canadian Electricity Exports: An Unholy Alliance?" Congressional Research Service, Library of Congress, September 15, 1982, p. i.
13 Ibid.
14 U.S. Congress, House Committee on Foreign Affairs, *United States–Canadian Relations and Acid Rain, Hearings before the Subcommittee on Human Rights and International Organizations and on Inter-American Affairs*, May 20, 1981.
15 Ibid., pp. 11–16.
16 Ibid., pp. 40–41.
17 Clarkson, *Canada and the Reagan Challenge*, p. 196.
18 "Acid Rain Meeting of Foreign Ministers," *International Perspectives* (April–May 1983): 4.
19 "President Reagan Assigns EPA Four Priority Tasks," *EPA Journal* (July

1983): 10.

20   "Ruckelshaus Outlines Major Issues," *EPA Journal* (July 1983): 8.

21   *Congressional Quarterly* (October 22, 1983): 2186.

22   U.S. Congress, Senate Committee on Environment and Public Works, *Hearings on S. 768, Re-authorization of the Clean Air Act*, February 2, 1984.

23   *Washington Post*, weekly national edition, December 12, 1983, p. 9.

24   Canadian Embassy, Washington, D.C., "Canada Responds to United States Inaction on Acid Rain," February 22, 1984.

25   Drew Lewis and William Davis, *Joint Report of the Special Envoys on Acid Rain*, January 1986, p. 5.

26   Ibid., p. 6.

27   Ibid., pp. 29–35.

28   Environment Canada, *Preliminary Assessment of Proposed U.S. Clean Coal Technology Research and Development and Demonstration Initiatives between 1986 and 1992*, Ottawa, January 1987.

29   Telephone interview with Rodney Bell, deputy director, Division of External Affairs, Ottawa, February 15, 1988.

30   Miles Godfrey, Canadian Consulate General Dallas, "Introduction to Environmental Effects of Acid Rain" (Address to the LBJ School of Public Affairs, University of Texas, April 18, 1984), pp. 25–26.

31   "Relations with the United States: Acid Rain," *International Perspectives, Supplement*, (January–February 1983): 4.

32   William Ruckelshaus, speech delivered on October 26, 1983.

33   "Relations with the United States: Acid Rain," p. 4.

34   John Roberts, minister of the environment, speech to the Air Pollution Control Association, New Orleans, June 21, 1982, p. 2.

35   "Relations with the United States: Acid Rain," pp. 3–4.

36   Canadian Embassy, Washington, D.C., "Canada Responds to United States Inaction on Acid Rain," February 22, 1984.

37   National Acid Rain Precipitation Assessment Program (NAPAP), *Interim Assessment: The Causes and Effects of Acidic Deposition*, vol. 1, *Executive Summary* (Washington, D.C.: Government Printing Office, 1987).

38   For further details see chapter 7.

39   Clarkson, *Canada and the Reagan Challenge*, p. 196.

40   For details, see chapter 5.

41   For a legal interpretation of Section 115 of the Clean Air Act, see chapter 9.

42   U.S. Clean Air Act, sec. 115.

Chapter Four

1   A. P. Altshuller and G. A. McBean, *The Long-Range Transport of Air Pollutants Problems in North America: A Preliminary Overview*, United States–Canada Bilateral Consultation Group, October 1979.

2   *Second Report of the United States–Canada Bilateral Research Consul-*

tation Group on the Long-Range Transport of Air Pollutants, October 1980.

3 Altshuller and McBean, *Long-Range Transport of Air Pollutants*, p. 24.

4 *Second Report on the Long-Range Transport of Air Pollutants*, p. 4.

5 John E. Carroll, *Acid Rain: An Issue in Canadian-American Relations* (Washington, D.C.: Canadian-American Committee, 1982), p. 39.

6 There was no Phase II report for Work Group 3B because they had no new information, nor was there a report from Work Group 3A, which had lost its research mandate.

7 The tables were compiled from information in United States–Canada, *Memorandum of Intent on Transboundary Air Pollution, Impact Assessment, Work Group I, Final Report*, January 1983, pp. I-1–I-24.

8 Interview with Lowell Smith, U.S. Environmental Protection Agency, April 1983.

9 United States–Canada, *Memorandum of Intent on Transboundary Air Pollution, Work Group 3B, Emissions, Costs, and Engineering Assessments*. Final Report, June 1982.

10 The members included: Bill Nierenberg (chairman), Ruth Patrick, David Evans, Bill Ackerman, Ken Rahn, Sherwood Rowland, Melvin Ruderman, Fred Singer, and John Roberts.

11 Gregory S. Wetstone and Armin Rosencranz, *Acid Rain in Europe and North America: National Responses to an International Problem* (Washington, D.C.: Environmental Law Institute, 1983), p. 128.

12 Eliot Marshall, "Air Pollution Clouds U.S.-Canadian Relations," *Science* (September 17, 1982): 1118.

13 Interview with Larry Regens, U.S. Environmental Protection Agency, January 1983.

14 Office of Science and Technology Policy, Executive Office of the President, news release, June 28, 1983. A summary of the panel's findings can be found in National Academy of Sciences, *NewsReport* 33, no. 6 (July–August 1983): 6–10.

15 Eliot Marshal, "Acid Rain, A Year Later," *Science* (July 15, 1983): 241.

16 Royal Society of Canada, "Report of the Peer Review Group" (1983), mimeographed.

17 ITF activities are funded from earmarked items in the budget of participating agencies.

18 For details on ITF activities, see chapter 7.

19 National Acid Precipitation Assessment Program, *Annual Report 1986* (Washington, D.C.: Government Printing Office, 1987), pp. 17–22.

20 National Acid Precipitation Assessment Program, *Annual Report 1982* (Washington, D.C.: Government Printing Office, 1983), p. 51.

21 Miles Godfrey, Canadian Consulate General Dallas, "Introduction to Environmental Effects of Acid Rain" (Address to the LBJ School of Public Affairs, April 18, 1984), p. 9.

22 Interview with Myron Uman, National Academy of Sciences, April 1983.

23 National Academy of Sciences, *Atmosphere-Biosphere Interactions:*

Toward a Better Understanding of the Ecological Consequences of Fossil Fuel Combustion (Washington, D.C.: National Academy Press, 1981), p. 7.

24 National Academy of Sciences, *Acid Deposition: Atmospheric Processes in Eastern North America* (Washington, D.C.: National Academy Press, June 1983).

25 Laval Lapointe, "Acid Precipitation Issue: Have We Lost Sight of Past Lessons?" *Proceedings of Wingspread Conference* (St. Paul, Minn.: Acid Rain Foundation, September 23–25, 1986), pp. 468–98.

26 For more information, see chapter 8.

Chapter Five

1 For more information, see chapter 2.

2 National Acid Precipitation Assessment Program, *Interim Asessment, The Causes and Effects of Acidic Deposition*, vol. 1, *Executive Summary* (Washington, D.C.: Government Printing Office, 1987), p. 1-10.

3 U.S. EPA, "New Source Performance Standards for Major Sulfur Dioxide Emitting Stationary Sources," *Federal Register* 37 (1972): 4767–69.

4 National Commission on Air Quality, *To Breathe Clean Air* (Washington, D.C.: Government Printing Office, 1981), sec. 3.9, p. 2. The report was commissioned in preparation for the reauthorization of the Clean Air Act as amended August 1977, P.L. 95-95.

5 Jeremy Hales, "Atmospheric Chemistry—A Lay Person's Introduction," in *Acid Rain: The Relationship between Sources and Receptors*, ed. James White (New York: Elsevier Science Publishing, 1988), pp. 71–89.

6 The use of the term *ethical* here is not to suggest that there is an "ethical" position and an "unethical" position. Rather, it identifies an argument in which the positions of the debaters are determined by values attached to each debater's particular situation. This usage is consistent with the definition of ethics from *Webster's Seventh Collegiate Dictionary*: "The rules of conduct recognized in respect to a particular class of human actions or a particular group, culture, etc."

7 The difficulties of the economic argument are indicated in the report by Larry B. Parker with Robert E. Trumble, "Proposed Acid Rain Legislation: Comparison of Cost Estimates For 10 Million Ton Reduction in SO_2 Emissions," Congressional Research Service, Library of Congress, April 19, 1982, mimeo. (Hereinafter cited as Parker, "Comparison of Cost Estimates.")

8 U.S. Congress, Office of Technology Assessment (OTA), *Acid Rain and Transported Air Pollutants: Implications for Public Policy* (Washington, D.C.: Government Printing Office, 1984), p. 49.

9 Drew Lewis and William Davis, *Joint Report of the Special Envoys on Acid Rain*, January 1986, p. 23.

10 Sierra Club, "A Stronger Clean Air Act: What Are the Costs?" *National News Report* 19 (November 19, 1987): 2.

11   OTA, *Acid Rain and Transported Air Pollutants*, p. 13.
12   Marchant Wentworth, "Acid Rain in Your Own Backyard," *Outdoor America* (Summer 1986): 12.
13   Lewis and Davis, *Joint Report*, p. 23.
14   Ibid., p. 14. Most of the proposed legislation aims to reduce SO_2 levels by 8–12 million tons, which, according to these estimates would cost $2–$6 billion.
15   U.S. Congress, House Committee on Energy and Commerce, Subcommittee on Health and Environment, *Hearings, Acid Deposition Control Act, H.R. 4567*, 99th Cong., 2d sess., May 1, 1986. Supplemental information on estimated cost increases for automobiles was provided by Ford ($27, p. 162), Volkswagen ($72, pp. 165–67), and Chrysler ($99, p. 172).
16   OTA, *Acid Rain and Transported Air Pollutants*, p. 141. The figure $1.5 billion represents the cost of installing scrubbers on the largest utility plants. These plants emitted about 7.6 million tons of SO_2 in 1980 and consumed about 60 percent of the high-sulfur coal.
17   Congressional Budget Office, *Curbing Acid Rain: Cost, Budget, and Coal-Market Effects* (Washington, D.C.: Government Printing Office, 1986), p. 6.
18   Ibid., p. xix.
19   David Streets, "The Design of Cost-effective Strategies to Control Acidic Deposition," in *The Acid Rain Debate: Scientific, Economic, and Political Dimensions*, ed. Ernest J. Yanarella and Randal H. Ihara (Boulder, Colo.: Westview Press, 1985), pp. 183–90.
20   For an analysis and rebuttal of the conspiracy argument, see Larry Parker, "Acid Rain Legislation and Canadian Electricity Exports: An Unholy Alliance?" U.S. Congressional Research Service, Library of Congress, September 15, 1982, mimeo. (Hereinafter cited as Parker, "Unholy Alliance?")
21   For details of legislative initiatives, see chapter 7.
22   NCAC includes these organizations: Amalgamated Clothing and Textile Workers; Americans for Democratic Action; American Lung Association; Center for Auto Safety; Citizens for a Better Environment; Environmental Action; Environmental Defense Fund; Environmental Policy Center; Environmentalists for Full Employment; Friends of the Earth; International Association of Machinists and Aerospace Workers; Izaak Walton League of America; League of Women Voters of the United States; National Association of Railway Passengers; National Audubon Society; National Consumers' League; National Farmers' Union; National Parks and Conservation Association; National Urban League; National Wildlife Federation; Natural Resources Defense Council; Oil, Chemical, and Atomic Workers' International Union; Sierra Club; United Steelworkers of America; Western Organization of Resource Councils; and The Wilderness Society.
23   National Clean Air Coalition, press release dated November 17, 1981. This release discusses: ICF, Inc., "Cost and Coal Production Effects of Reducing Electric Utility Sulfur Dioxide Emissions," prepared for the National Wildlife Federation and the NCAC, November 14, 1981.) (Herein-

after cited as ICF-NWF study.)

24 U.S. Congress, Senate Committee on Energy and Natural Resources, *Hearings, Acid Precipitation and the Use of Fossil Fuels*, 1982. (Hereinafter cited as Senate, *Hearings.*) The current reference is to the testimony of Michael Oppenheimer, senior scientist, Environmental Defense Fund, pp. 1229–48.

25 Robert M. Wolcott and Adam Z. Rose, "The Economic Effects of the Clean Air Act," prepared for the Natural Resources Defense Council, Public Interest Economics Foundation, Washington, D.C., March 1982. For comparison, the U.S. Office of Technology Assessment has estimated that controls would cost between $2.5 and $4.7 billion a year by 1990. Canadian control costs have been estimated to range from $600 million to $1 billion.

26 Among these are Ellis B. Cowling, "An Historical Resumé of Progress in Scientific and Public Understanding of Acid Precipitation and Its Biological Consequences," SNSF Project Research Report Series (Oslo-Aas, Norway: SNSF Project, 1981); Eville Gorham, "Ecological Aspects of the Chemistry of Atmospheric Precipitation," in *Multidisciplinary Research Related to Atmospheric Sciences*, ed M. H. Glantz, H. van Loon, and E. Armstrong (Boulder, Colo.: National Center for Atmospheric Research, 1978), pp. 256–96; G. E. Likens, F. H. Bormann, and N. M. Johnson, "Acid Rain," *Environment* 14 (1972): 33–40.

27 ICF-NWF study, p. 16.

28 Natural Resources Defense Council, "Why AEP Cost Estimates Are So High," press release, September 1982.

29 Discussed in Parker, "Unholy Alliance?"

30 NCAC, "Acid Rain Control Costs," press release, December 1982.

31 Interview with Edwin Anthony, NCA, January 1983.

32 Rochelle Stanfield, "The Acid Rainmakers," *National Journal* (June 14, 1986): 1500–1503.

33 Science and Policy Associates, Inc., *Acidic Deposition Catalog* (Palo Alto, Calif.: Electric Power Research Institute, August 1986).

34 Ibid.

35 U.S. Congress, House Committee on Energy and Commerce, Subcommittee on Health and Environment, *Hearings, Acid Precipitation: Effects and Solutions to Combat Acid Precipitation*, 97th Cong., 1st sess., 1981. (Hereinafter cited as House, *Hearings.*) Dr. Brocksen is program manager of ecological studies, Environmental Assessment Department, Edison Power Research Institute.

36 Ibid.

37 Ibid.

38 House, *Hearings*. Mr. Courtney is director of air quality for Commonwealth Edison Co.

39 Ibid.

40 NAS, *NewsReport* (July–August 1983): 7.

41 Sierra Club, "Acid Rain: Congress Moves on Controls," *National News*

Report 18 (May 27, 1986).

42 EEI, "Acid Rain and the Clean Air Act," issue summary, January 1988.

43 Peabody Coal Company, "Economic Impacts of Senate Environment and Public Works Committee Acid Rain Amendment," fact sheet, August 1982.

44 Parker, "Comparison of Cost Estimates."

45 Interview with Edwin Anthony, NCA, January 1983.

46 Carl E. Bagge, "Acid Rain—Canada's Rights and Remedies" (Address to the Sixty-fourth Annual Meeting of the Canadian Bar Association, August 30, 1982).

47 U.S. Congress, Senate Committee on Environment and Public Works, Subcommittee on Environmental Protection, *Hearings, Clean Air Act Amendments of 1987*, 100th Cong., 1st sess., 1987, pp. 637–48.

48 Parker, "Comparison of Cost Estimates."

49 Ibid.

50 Sierra Club, "Congressional Recess Begins: Acid Rain Top Priority," *National News Report* 18 (August 26, 1986): 1–2.

51 Confirmation of this reason comes from several sources. In interviews Martha Kettel of TVA, Edwin Anthony of NCA, and William McGonnell of EEI all suggested that Freeman's influence was important in TVA's position. This influence is given its most negative form in a comment from *Electrical Week*, April 5, 1982, p. 5: "[TVA] staffers blame director S. David Freeman for leading the board to go further than it had to on the issue and for ignoring competent staff advice on the subject."

52 House, *Hearings*, pp. 139–54; and TVA, "Acid Precipitation Legislation," position paper released in March 1982.

53 TVA, "Acid Precipitation Legislation," p. 3.

54 TVA, "Impact on TVA of the Various Clean Air and Acid Rain Bills Pending in Congress," position paper, n.d., p. 1.

55 TVA, "Impact on TVA," p. 2, estimates 8–13 percent increases based on S 1706 or S 1709 A letter from C. H. Dean, Jr., chairman of the TVA board, to William B. Sturgill of the Kentucky Energy Cabinet states in reference to S 3041: "A rough analysis of this bill indicated that TVA would incur increases in costs of between 3 and 6 percent."

56 "AEP, TVA Split on Acid Rain Issue," *Electrical Week* (March 22, 1982): 2.

57 TVA, "Acid Precipitation Legislation," p. 4.

58 Ned Helme and Chris Neme, *Acid Rain: Road to a Middleground Solution* (Washington, D.C.: Center for Clean Air Policy, 1987).

59 R. J. Beamish and Harold Harvey, "Acidification of the La Cloche Mountain Lakes, Ontario, and Resulting Fish Mortalities," *Journal of Fisheries Research Board of Canada* 29 (1972): 1131–43.

60 Interview with Adele Hurley, co-director, Canadian Coalition on Acid Rain, April 1983.

61 CCAR member groups include: Algonquin Wildlands League; Allied Boating Association of Canada; Association Quebecoise de Technique de l'Eau; Canadian Environmental Law Association; Canadian Society of Environmental Biologists; Canadian Nature Federation; Canadian Fishing Tackle

Industry; Canadian Sport Fishing Institute; Canadian Sporting Goods Association; Conservation Council of Ontario; Consumers' Association of Canada—Manitoba; Federation of Ontario Cottagers' Association; Federation of Ontario Naturalists; International Atlantic Salmon Foundation; Inuit Tapirisat of Canada; Muskies Canada; Muskoka Lakes Association; National Survival Institute; Natural History Society of P.E.I.; Northern Ontario Tourist Outfitters' Association; Ontario Federation of Anglers and Hunters; Ontario Forestry Association; Ontario Log Builders' Association; Ontario Lung Association; Ontario Public Interest Research Group; Ontario Steelhead and Salmon Fishermen; Pollution Probe—Ottawa; Pollution Probe—Toronto; Resorts Ontario; Save Our Streams; Sierra Club of Ontario; Societé Pour Vaincre la Pollution; STOP (Montreal); Temagami Lakes Association; Toronto Sportsmen's Association; Tourism Industry Association of Canada; Tourism Ontario; Union of Ontario Indians; USC Social Research Commission—University of Western Ontario; Waterloo Public Interest Research Group; Watson Lakes Trust; West Coast Environmental Law Association.

62 Interview with Michael Perley, co-director, CCAR, April 1983.
63 Interview with Adele Hurley, co-director, CCAR, April 1983. For details of this action taken by the provincial government of Ontario, see chapter 7.
64 Interview with Adele Hurley, April 1983.
65 Interview with Michael Perley, April 1983.
66 For more detailed accounts of INCO's economic situation and emissions reduction efforts, see transcripts of the following: Dr. Walter Curlook, executive vice president, INCO, Ltd., presentation to the Federal Subcommittee on Acid Rain, Ottawa, Ontario, June 22, 1983; and W. Charles Ferguson, director, government programs, INCO, Ltd., Testimony before the New England Congressional Caucus Hearing, Concord, New Hampshire, April 26, 1982.
67 Interview with W. Charles Ferguson, INCO, January 1983.
68 Environment Ontario, *Summary and Analysis of the Third Progress Reports (July 31, 1987) by Ontario's Four Major Emitters of Sulphur Dioxide*.
69 Interview with Ferguson, INCO, Ltd., January 1983.
70 Interview with Howard Marks, legislative aide to Senator Charles Percy, January 1983.
71 Ontario Hydro, "Acid Gas Control Program."
72 Parker, "Unholy Alliance?"
73 Ibid.
74 National Wildlife Federation, *Conservation 83* 1, no. 6 (April 29, 1983): 5.
75 Interview with William McGonnell, EEI, January 1983.
76 Unpublished transcript of a workshop held at the LBJ School of Public Affairs, University of Texas at Austin, April 28–29, 1983.
77 Rochelle Stanfield, "The Acid Rainmakers," pp. 1500–1503.
78 Unpublished transcript.

Chapter Six

1   John Roberts, "Acid Rain: A Serious Bilateral Issue" (Address to the Air Pollution Control Association, June 21, 1982), Canada, Department of External Affairs, *Statements and Speeches* 82/11.

2   Canadian Embassy, "How Many More Lakes Have to Die?" *Canada Today* 12, no. 2 (February 1981): 3. A bilateral research group working under the Memorandum of Understanding reported in 1984 that enough reliable monitoring information was now available to say with confidence that half of the SO_2 deposited in Canada is imported from the United States; see Miles Godfrey, "Introduction to Environmental Effects of Acid Rain" (Address to the LBJ School of Public Affairs, April 18, 1984).

3   John Roberts, "The Urgency of Controlling Acid Rain" (Address to the Air Pollution Control Association, Montreal, June 23, 1980), Canada, Department of External Affairs, *Statements and Speeches* 80/8; and Canada, House of Commons, Committee on Fisheries and Forestry, Subcommittee on Acid Rain, *Still Waters: The Chilling Reality of Acid Rain* (Ottawa: Minister of Supplies and Services, 1981), p. 11.

4   Margaret Munro, "The Silent Peril," *Ottawa Citizen* (reprint, Environment Canada, 1981), p. 2.

5   Ibid., p. 1, and Canada, House of Commons, *Still Waters*, p. 12.

6   *New York Times*, February 10, 1984.

7   Lois R. Ember, "Acid Pollutants: Hitchhikers Ride the Wind," *Chemical and Engineering News* 59, no. 37 (September 14, 1981): 21.

8   In this chapter references to the Clean Air Act without further qualification are to the Canadian statute.

9   *Statutes of Canada*, Clean Air Act, 28th Parliament, 23rd sess., vol. 1 (1971–72–73), sec. 3.

10   Ibid., sec. 4.

11   Gregory Wetstone and Armin Rosencranz, *Acid Rain in Europe and North America: National Responses to an International Problem* (Washington, D.C.: Environmental Law Institute, 1983), p. 146.

12   David Estrin and John Swaigen, *Environment on Trial—A Handbook of Ontario Environmental Law*, rev. ed. (Toronto: Canadian Environmental Law Research Foundation, 1978), p. 92.

13   Drew Lewis and William Davis, *Joint Report of the Special Envoys on Acid Rain*, January 1986, p. 24.

14   Ibid., p. 82.

15   Ibid., p. 81.

16   Clean Air Act, sec. 8.

17   Canada–United States, *Memorandum of Intent on Transboundary Air Pollution, Report of Legal, Institutional, and Drafting Work Group*, July 31, 1981, p. 16.

18   Environment Canada, "Acid Rain: Control of Acid Gas Emissions," Air Pollution Control Directorate, Environmental Protection Service, 1982; and Canada, House of Commons, *Still Waters*, p. 15.

19 Canada–United States, *Memorandum of Intent*, p. 16.

20 Canada, House of Commons, *Still Waters*, p. 81. This problem is similar in the United States, where the U.S. Clean Air Act regulates old sources less stringently than new ones.

21 Clean Air Act, sec. 7.

22 Estrin and Swaigen, *Environment on Trial*, pp. 92–93.

23 Canada, House of Commons, *Minutes of Proceedings and Evidence of the Standing Committee on Fisheries and Forestry*, 11 August 1978, p. 5:23; and Wetstone and Rosencranz, *Acid Rain in Europe and North America*, p. 119.

24 Canada–United States, *Memorandum of Intent, Report of Legal, Institutional, and Drafting Work Group*, pp. 16–17.

25 Canada, House of Commons, *Still Waters*, p. 83.

26 Clean Air Act Amendment, Bill C-51, 1980.

27 Canada–United States, *Memorandum of Intent*, pp. 14–15.

28 Canada, House of Commons, *Debates*, 16 December 1980, pp. 5800–5801.

29 Under U.S. law the Clean Air Act is implemented by state implementation plans, which are developed by the states and approved by the federal government.

30 Canada, House of Commons, *Still Waters*, p. 47.

31 Michael Perley, "Acid Rain: How Far Have We Come?" *Probe Post* (Spring 1986): 22–24.

32 Information from the Canadian embassy, Washington, D.C.

33 Environment Canada, "Acid Rain: What It Is and What It Does," pamphlet.

34 Ibid.

35 Canada, House of Commons, *Minutes of the Proceedings and Evidence of the Standing Committee on Fisheries and Forestry*, June 10, 1980, p. 3.12.

36 Ibid., November 13, 1979, p. 6.26.

37 Environment Canada, "Acid Rain: A Chronology of Events," Air Pollution Directorate, Environmental Protection Service, Ottawa, August 1982, mimeo, p. 2.

38 Ontario Ministry of the Environment, "The Case against Acid Rain: A Report on Acidic Precipitation and Ontario Programs for Remedial Action," October 1980, rev. 1982, p. F.

39 Environment Canada, "Stopping Acid Rain," pamphlet.

40 Drew Lewis and William Davis, *Joint Report of the Special Envoys on Acid Rain*, January 1986.

41 Mark MacGuigan, "Acid Rain: One of the Most Serious Problems in Canada-U.S. Relations" (Address to the Conference on Acid Rain, May 2, 1981), Canada, Department of External Affairs, *Statements and Speeches*, 81/10.

42 For example, see U.S. Congress, House Committee on Science and Technology, *Acid Rain: Hearings before the Subcommittee on Natural Resources, Agriculture Research, and Environment*, 97th Cong., 1st sess., September, 18, 19, November 19, December 9, 1981.

43   Environment Canada, "Long Range Transport of Air Pollution/Acid Rain Programs—Status Summary," Air Pollution Control Directorate, Environmental Protection Service, Ottawa, January 1982, mimeo, p. 3.

44   Ember, "Acid Pollutants," p. 30.

45   MacGuigan, "Acid Rain."

46   Environment Canada, "Acid Rain: Control of Acid Gas Emissions," p. 1.

47   John Roberts, "The Urgent Need to Control Acid Rain" (Address to the Georgia Conservancy League, June 24, 1982), Canada, Department of External Affairs, *Statements and Speeches*, 82/14.

48   For more details, see chapter 3.

49   James Regens and Robert Rycroft, "Perspectives on Acid Deposition Control: Science, Economics, and Policymaking," in *The Acid Rain Debate: Scientific, Economic, and Political Dimensions*, ed. Ernest Yanarella and Randal Ihara (Boulder, Colo.: Westview Press, 1985), p. 89.

50   Environmental Protection Act, *Revised Statutes of Ontario, 1980*, chap. 141, as amended by 1981, chap. 49, January 1982, pt. I, sec. 3.

51   Environmental Quality Act of 1972, *Revised Statutes of Quebec, 1977*, chap. Q-2, sec. 2.

52   Specifically, this is provided in Section 8 of the Ontario Environmental Protection Act and in Sections 20 and 22 of Quebec's Environmental Quality Act. The Quebec law further requires an operator of a plant to obtain a certificate of authorization in order to operate or use a process which releases contaminants into the environment.

53   In Ontario the director is given this authority if the emissions would pose "a hazard to the health or safety of any person, or impair the quality of the natural environment for any use that can be made of it" (Environmental Protection Act, Section 8). In Quebec similar authority is granted if the emissions are likely to "affect the life, health, safety, welfare, or comfort of human beings or to cause damage or otherwise impair the quality of the soil, vegetation, wildlife, or property" (Environmental Quality Act, Sections 20 and 22).

54   Environmental Protection Act, sec. 121; see David Estrin, "The Legal and Administrative Management of Ontario's Air Resources 1967—74," in *Environmental Management and Public Participation*, ed. P. S. Elder (Canada: Canadian Environmental Law Research Foundation and the Canadian Environmental Law Association, 1974), p. 182.

55   Environmental Quality Act, sec. 96.

56   Estrin and Swaigen, *Environment on Trial*, p. 33.

57   Environmental Protection Act, sec. 1.

58   This is provided in Section 6 of the Ontario Environmental Protection Act and in Section 25 of the Environmental Quality Act. Section 6 specifically includes damage to plant or animal life as an acceptable reason for issuing a control order. Therefore, the term "environment" used here means more than just the soil, air, and water within the province of Ontario (as it was narrowly defined for the certificate of approval/authorization).

59   Estrin and Swaigen, *Environment on Trial*, p. 35.

60   Ontario/Canada Task Force, "Report of the Ontario/Canada Task Force for the Development of Air Pollution Abatement Options for INCO, Limited, and Falconbridge Nickel Mines, Limited, in the Regional Municipality of Sudbury, Ontario," Fall 1982, p. 257.

61   The Ontario Environmental Protection Act states that: "The Lieutenant Governor in Council may make regulations prohibiting or regulating and controlling the depositing, addition, emission or discharge of any contaminant or contaminants into the natural environment from any source of contaminant or any class thereof" (Environmental Protection Act, sec. 136[1][b]). The Quebec Environmental Quality Act provides that: "The Government may make regulations to prohibit, limit, and control sources of contamination as well as the emission, deposit, issuance, or discharge into the environment throughout all or part of the territory of Quebec" (Environmental Quality Act, sec. 31[c]).

62   In Ontario a regulation is not subject to statutory appeal, and it is subject to challenge in the courts only on a limited range of issues. See Canada—United States, *Memorandum of Intent on Transboundary Air Pollution*, p. 21. The Quebec Environmental Quality Act does not provide for avenues of appeal to the Commission Municipale du Quebec for decisions made by the lieutenant governor in council.

63   This is found in Section 136(1)(b) of the Ontario Environmental Protection Act and in Section 31(d) of the Quebec Environmental Quality Act.

64   U.S. Congress, House Committee on Energy and Commerce, *Acid Precipitation: Hearings before the Subcommittee on Health and the Environment*, 97th Cong., 1st sess., October 1, 2, and 6, 1981, p. 553; and telephone conversation with Mr. John Martin, Legal Branch, Ontario Ministry of the Environment, May 16, 1983.

65   Acidic Precipitation in Ontario Study, *Annual Program Report Fiscal Year 1986/1987*, APIOS Report no. 005, APIOS Coordination Office, Ontario Ministry of the Environment, July 1987.

66   Marcel Leger, *Quebec and the Problem of Acid Precipitation* (Quebec Ministère de l'Environnement, 1981), p. 3.

67   Quebec Ministère de l'Environnement, "A Submission to the United States Environmental Protection Agency Opposing Relaxing of SO_2 Emission Limits in State Implementation Plans and Urging Enforcement," September 11, 1981, pp. 10–16. Also, Quebec Ministère de l'Environnement, "Dossier sur les Precipitations Acides," Conference des Premiers Ministres des Provinces de l'Est et des Gouverneurs des Etats du Nord-Est Americain, Rockport, Maine, June 21, 1981, p. 6.

68   Ontario Ministry of the Environment, "The Case against Acid Rain," p. D.

69   Gregory Wetstone and Armin Rosencranz, *Acid Rain in Europe and North America*, p. 115.

70   INCO, uncontrolled, would emit 7,000 tons of SO_2 per day. In 1970, in order to reduce local air pollution, INCO was issued a control order to reduce emissions below 3,600 tons per day by 1976, and to 750 tons per

day by 1979. INCO requested the Ministry of the Environment to reevaluate the control order to reduce emissions below 3,600 tons per day, and the ministry rescinded the order. See Robert Paehkle, "Overview —Canada," *Environment Magazine*, September 1979.

71   Canadian Embassy, *Aide Memoire*, Washington, D.C., March 6, 1981, p. 2.
72   Keith Norton, in *Acid Rain: A Transjurisdictional Problem in Search of a Solution, Proceedings of a Conference*, ed. Peter S. Gold (Buffalo, N.Y.: Canadian-American Center, State University of New York at Buffalo, 1982), p. 59.
73   U.S. Congress, House Committee on Science and Technology, *Acid Rain: Hearings before the Subcommittee on Natural Resources, Agriculture Research, and Environment.*
74   Ontario Ministry of the Environment, "The Case against Acid Rain," p. D; and Canadian Embassy, *Acid Rain*, and "Ontario Hydro Fact Sheet," 1980.
75   Ministry of the Environment, "Ontario's Acid Gas Control Program for 1986–1994," pamphlet, Ontario.
76   Ibid.
77   Quebec Ministère de l'Environnement, "Dossier sur les Precipitations Acides," p. 41.
78   Canada, House of Commons, *Still Waters*, p. 24.
79   Peter Dunn, conseilleur, Quebec Government House, personal correspondence with the author, March 16, 1983.
80   Quebec Ministère de l'Environnement, "A Submission to the United States Environmental Protection Agency," p. 37.
81   *Acid Rain 1986: A Handbook for States and Provinces, Proceedings of Wingspread Conference* (St. Paul, Minn.: Acid Rain Foundation, September 23–25, 1986), p. 480.
82   Environment Canada, "Acid Rain: What It Is and What It Does," pamphlet.
83   Ontario Ministry of the Environment, "The Case against Acid Rain," p. C; and Ontario Ministry of the Environment, "Ontario Takes Battle to United States Courts," *Legacy* 10, no. 1 (May–June 1981): 1.
84   Quebec Ministère de l'Environnement, "A Submission to the United States Environmental Protection Agency."
85   Ontario Ministry of the Environment, "The Case against Acid Rain," p. C.
86   *New York Times*, March 21, 1984.
87   Environment Canada, "Acid Rain: Control of Acid Gas Emissions," Air Pollution Control Directorate, Environmental Protection Service, 1982.
88   U.S. Congress, House Committee on Energy and Commerce, *Hearings*, October 1, 2, and 6, 1981.
89   Ontario Ministry of the Environment, "The Case against Acid Rain," p. C.
90   Leger, *Quebec and the Problem of Acid Precipitation*, pp. 1–2; and "New York–Quebec: Agreement on Acid Precipitation," *International Legal Materials* 21, no. 4 (July 1981): 721–25.
91   Leger, *Quebec and the Problem of Acid Precipitation*, p. 1.

Chapter Seven

1 For an account of the bilateral negotiations, see chapter 3.
2 The information in this section depends heavily upon the National Commission on Air Quality report, *To Breathe Clean Air* (Washington, D.C.: Government Printing Office, 1981), sec. 3.9, p. 2. The report was commissioned in preparation for the reauthorization of the Clean Air Act as amended in August 1977, P.L. 95-95.
3 Sierra Club, "Clean Air Standards before the Courts," *National News Report* 17 (December 19, 1985): 1.
4 Motor Vehicle Control Act of 1965, P.L. 89-272.
5 Interview with Victor Mearki, aide to Senator Stafford, Senate Committee on Environment and Public Works, January 13, 1983.
6 National Commission on Air Quality, *To Breathe Clean Air*, sec. 3.9, p. 24.
7 An attempt to do so was made toward the end of the Carter administration. For details, see chapters 6 and 9.
8 Eliot Marshall, "Air Pollution Clouds U.S.-Canadian Relations," *Science* (September 17, 1982): 1118–19.
9 National Commission on Air Quality, *To Breathe Clean Air*, p. ix.
10 Lennart Lundvist, *The Hare and the Tortoise* (Ann Arbor: University of Michigan Press, 1980).
11 Ellis B. Cowling, "Acid Precipitation in a Historical Perspective," *Environmental Science and Technology* 2 (1982): 110–22.
12 National Acid Precipitation Assessment Program, *Annual Report, 1986* (Washington, D.C.: Government Printing Office, 1987), p. 25.
13 Interagency Task Force on Acid Precipitation, "National Acid Deposition Assessment Plan, Annual Report 1984" (Washington, D.C.: EPO Publications, 1985), pp. 11–13.
14 Kathleen Bennett, "Testimony before the U.S. Senate Committee on Energy and Natural Resources," *Hearings on Acid Precipitation and the Use of Fossil Fuel*, 19 August 1982.
15 National Acid Precipitation Assessment Program, *Annual Report, 1982* (Washington, D.C.: Government Printing Office, 1983), pp. 1–2.
16 Lawrence Mosher, "Acid Rain Debate May Play a Role in the 1984 Presidential Sweepstakes," *National Journal* (May 14, 1983): 998–99.
17 Lawrence Mosher, "Administration Loses Its Umbrella against Standfast Acid Rain Policy," *National Journal* (July 30, 1983): 1590–91.
18 Rochelle Stanfield, "Regional Tensions Complicate Search for an Acid Rain Remedy," *National Journal* (May 5, 1984): 860–63.
19 Telephone conversation with Alex Cristofaro, air pollution policy analyst, EPA, February 1988.
20 National Acid Rain Precipitation Assessment Program (NAPAP) *Interim Assessment: The Causes and Effects of Acidic Deposition*, vol. 1, *Executive Summary* (Washington, D.C.: Government Printing Office, 1987).
21 Interview with Victor Mearki, January 13, 1983.
22 Louis Harris, "Testimony before the U.S. Senate Committee on Environ-

ment and Public Works," *Hearings on the Clean Air Act*, October 1981.

23 The National Clean Air Coalition is discussed in chapter 5.

24 National Clean Air Coalition, "Positions on the Clean Air Act" (April 3, 1981), p. 23.

25 *Inside EPA* (June 5, 1982): 95–96.

26 U.S. Senate, Clean Air Act Amendments of 1982, S 3041, 1982.

27 U.S. Congress, Senate Committee on Environment and Public Works, *Report on the 1982 Clean Air Act Amendments to Accompany S 3041*, November 15, 1982, p. 203.

28 Committee Report, p. 124.

29 Committee Report, p. 109.

30 Kathleen Bennett, "Testimony before the Senate Committee on Energy and Natural Resources," August 19, 1982.

31 "Senate Committee May Develop Air Act Draft in Effort to Reach an Agreement," *Inside EPA* (February 15, 1982): 3.

32 Rochelle Stanfield, "The Acid Rainmakers," *National Journal* (June 14, 1986): 1500–1503.

33 Sierra Club, "Acid Rain: Congress Moves on Controls," *National News Report* 18 (May 27, 1986): 1.

34 U.S. Congress, Senate Committee on Environment and Public Works, *Report on the Clean Air Standards Attainment Act of 1987*, November 19, 1987, p. 388.

35 Sierra Club, "Waxman Subcommittee to Hold Clean Air Mark-ups," *National News Report* 20 (February 5, 1988): 1.

36 Betty Brink, "Coalition Pushing to Clean up Clean Air Act," *Texas Observer*, January 15, 1988.

37 Rochelle Stanfield, "Energy Focus," *National Journal* (March 8, 1986): 612.

38 Chapter 590 of the Acts of 1985, Commonwealth of Massachusetts.

39 Chapter 125-D, Acid Rain Control Act, New Hampshire, 1985.

40 New York State Department of Environmental Conservation, "The Sulfur Deposition Control Program," pamphlet, June 1985.

41 Minnesota Pollution Control Agency, *Acid Precipitation Program*, 1986 Biennial Report to the Legislature.

42 Wisconsin Department of Natural Resources, *Wisconsin Water Quality, 1986 Report to Congress*, Madison, Wisc., PUBL-WR-137 86, p. 56.

43 State of Michigan, "Sulfur Dioxide Regulations," mimeo.

44 Most proposals are limited to taxing emissions of SO_2. A comprehensive tax bill covering SO_2 and NO_x emissions was introduced in 1983 by Senator David Durenberger of Minnesota. S 2001 would yield $40 billion over a period of ten years, two-thirds from SO_2 and one-third from NO_x sources. A tax on electric utility rates has been proposed by Governor John H. Sununu, who was chairman of the National Governors' Association Acid Rain Task Force. See John H. Sununu, "Acid Rain: Sharing the Cost," *Issues in Science and Technology* 1 (Winter 1985): 47–59.

Chapter Eight

1   Raymond M. Robinson, "Notes For an Address at the Seventh Symposium on Statistics and the Environment" (Washington, D.C.: National Academy of Sciences, 1982), mimeo, p. 18.

2   C. J. Chacko, *The International Joint Commission* (New York: Columbia University Press, 1932), p. 21. For the text of the Boundary Waters Treaty, see U.S. 36 Stat. 2448, 1911.

3   Preamble to the Boundary Waters Treaty of 1909.

4   O. P. Dwivedi and John E. Carroll, "Issues in Canadian-American Environmental Relations," in *Resources and the Environment: Policy Perspectives for Canada*, ed. O. P. Dwivedi (Toronto: McClelland and Stewart, 1980), p. 311.

5   Robinson, "Seventh Symposium on Statistics and the Environment," pp. 6–7.

6   International Joint Commission, *Seventy Years of Accomplishment, 1979 Annual Report* (Ottawa and Washington, D.C.: IJC, 1980), p. 53. (Hereinafter cited as IJC, *1979 Annual Report*.)

7   Charles R. Ross, "The IJC: A U.S. View," in *Toward a Better Understanding of Canadian-American Relations*, ed. A. P. Splete (Canton, N.Y.: St. Lawrence University), p. 71.

8   Based on General Accounting Office, "Cleaning up the Great Lakes: United States and Canada are Making Progress in Controlling Pollution from Cities and Towns," RED-75-338, March 21, 1975, p. 1.

9   Boundary Waters Treaty of 1909, Article 10.

10 International Joint Commission, "Summary of the Seminar on the IJC: Its Achievements, Needs, and Potential," Montreal, 1974, mimeo, p. 16.

11 "IJC Fact Sheet," April 1982.

12 David G. Marquand and Anthony Scott, "Canada's International Environmental Relations," in *Resources and the Environment: Policy Perspectives for Canada*, ed. O. P. Dwivedi (Toronto: McClelland and Stewart, 1980), p. 79.

13 Robinson, "Seventh Symposium on Statistics and the Environment," p. 7.

14 International Joint Commission, *Transboundary Air Pollution: Detroit and St. Clair River Areas, 1972 Report* (Washington D.C.: Government Printing Office, 1973) p. 1.

15 Confidential interview, IJC, January 1983.

16 General Accounting Office, "IJC Water Quality Activities," p. 9.

17 Confidential interview, IJC, January 1983.

18 Ibid.

19 Confidential interview, IJC, January 1983. See also the discussion by Don Munton, "Paradoxes and Prospects," in *The International Joint Commission Seventy Years On*, ed. Robert Spencer (Toronto: University of Toronto Press, 1981), pp. 75–81.

20 Confidential interview, IJC, January 1983.

21 Leonard B. Dworskey, George R. Francis, and Charles Swezey, "Manage-

ment of the International Great Lakes," *Natural Resources Journal* 14 (January 1974): 103–4; and U.S. Government Accounting Office, "A More Comprehensive Approach Is Needed to Clean up the Great Lakes," CED-82-63, May 21, 1982, p. 1.

22 GAO, "Great Lakes," p. 1.

23 Carl A. Esterhay, "Restoring the Water Quality of the Great Lakes: The Joint Committee of Canada and the U.S.," *Canada–United States Law Journal* (Summer 1981): 218.

24 Dworskey et al., "Management of the International Great Lakes," pp. 21–23.

25 Ibid.

26 Fred E. Moseley, "The United States–Canadian Great Lakes Pollution Control Agreement: A Study in International Water Pollution Control" (Ph.D. diss., Kent State University, 1982), pp. 69–70.

27 Ibid. The three succeeding references were: 1955, St. Croix River; 1959, Rainy River and Lake of the Woods; and 1964, Red River.

28 See, e.g., Esterhay, "Restoring Water Quality."

29 Raymond M. Robinson, "Notes for an Address at the Seventh Symposium on Statistics and the Environment" (Washington, D.C.: National Academy of Sciences, 1981), mimeo, pp. 13–14; and International Joint Commission, *New and Revised Great Lakes Water Quality Objectives* (Washington, D.C.: Government Printing Office, May 1977), 1:8–9.

30 Kenneth Wardroper, "Canada's Interests as Regards Protection and Regulation of the Great Lakes," *Syracuse International Law Journal* 1 (1973): 209.

31 Moseley, "Pollution Control Agreement," p. 73.

32 Richard B. Bilder, "Controlling Great Lakes Pollution: A Study of U.S.-Canadian Cooperation," *Michigan Law Review* 70 (January 1972): 499–500.

33 Moseley, "Pollution Control Agreement," pp. 75–76.

34 Ibid., pp. 77, 101.

35 Most of the discussion that follows is based on Don Munton, "Great Lakes Water Quality: A Study in Environmental Politics and Diplomacy," in *Resources and the Environment: Policy Perspectives for Canada*, ed. O. P. Dwivedi (Toronto: McClelland and Stewart, 1980), pp. 159–64.

36 Munton, "Water Quality," p. 159; and Wardroper, "Canada's Interests," p. 208.

37 Bilder, "Controlling Pollution," p. 528.

38 Great Lakes Water Quality Agreement between the United States and Canada (1972), T.I.A.S. no. 7312, Article 5.

39 Munton, "Water Quality," pp. 159–64.

40 GAO, "Cleaning up the Great Lakes," p. 1.

41 Great Lakes Water Quality Agreement of 1978, Article 1.

42 Ibid., Annexes 1 and 10.

43 IJC, *1979 Annual Report*, p. 44.

44 The Trail smelter case is discussed in greater detail in chapter 9.

45 This 1972 event is discussed in IJC, memorandum, July 3, 1980.
46 International Joint Commission, *First Annual Report on Ontario-Michigan Air Pollution* (Ottawa and Washington, D.C.: Government Printing Office, 1976), p. 3.
47 Ibid.
48 IJC, memorandum, October 31, 1974.
49 IJC, memorandum, March 1975.
50 IJC, memorandum, July 3, 1980.
51 IJC, memorandum, October 24, 1978.
52 Don Munton, "Acid Rain and Basic Politics," *Alternatives* 10 (1981): 28.
53 Don Munton, "Dependence and Interdependence in Transboundary Environmental Relations," *International Journal* 36 (1980– 81): 165.
54 John E. Carroll, *Acid Rain: An Issue in Canadian-American Relations* (Toronto: C.D. Howe Institute, 1982), p. 39.
55 International Joint Commission, *Third Annual Report on Great Lakes Water Quality* (Ottawa and Washington, D.C.: Government Printing Office, 1974), p. 13.
56 IJC, memorandum, July 3, 1980.
57 IJC, memorandum, March 1, 1977.
58 IJC, memorandum, July 3, 1980.
59 For more information, see chapter 3.
60 Don Munton, "Paradoxes and Prospects," pp. 76–84.
61 International Joint Commission, *Sixth Annual Report on Great Lakes Water Quality* (Ottawa and Washington, D.C.: Government Printing Office, 1974), p. 21.
62 "Threat from Acid Rain Worsens," *Minneapolis Star*, July 17, 1979.
63 U.S. EPA, news release, January 16, 1981. For more details, see chapter 9.
64 See IJC, *1980 Annual Report* (Ottawa and Washington D.C.: Government Printing Office, 1981), p. 18.
65 IJC, *First Biennial Report under the Great Lakes Water Quality Agreement* (Ottawa and Washington D.C.: Government Printing Office, 1982), p. 20.
66 Telephone interview with Dr. Fisher, IJC, February 19, 1988.
67 Confidential interview, IJC, January 1983.
68 See especially Carroll, *Acid Rain: An Issue in Canadian-American Relations.*
69 Confidential interview, IJC, January 1983.

Chapter Nine

1 See chapter 8.
2 J. Alan Beesley, "The Canadian Approach to International Environmental Law," *Canadian Year Book of International Law* 11 (1973): 7.
3 John Roberts, "The Transnational Implications of Acid Rain," *Canada–United States Law Journal* 5 (1982): 2.
4 David Wooley, "Current Regulatory Framework: Introductory Comment," *Canada–United States Law Journal* 5 (1982): 60. For text of the Bound-

ary Waters Treaty, see U.S. 36 Stat. 2448, 1911.

5 George Alexandrowicz, "Closing Remarks," in *Acid Rain: A Transjuris-dictional Problem in Search of Solution, Proceedings of a Conference*, ed. Peter S. Gold (Buffalo N.Y.: Canadian-American Center Publications, State University of New York at Buffalo, 1982), p. 161.

6 Kenneth B. Hoffman, "State Responsibility in International Law and Transboundary Pollution Injuries," *International and Comparative Law Quarterly* 25 (1976): 511. Cross-jurisdictional lawsuits are not well suited to injunctive actions because of the difficulty in obtaining enforcement in other jurisdictions' courts. See "Private Remedies for Transboundary Injury in Canada and the United States: Constraints upon Having to Sue Where You Can Collect," *Ottawa Law Review* 10 (1978): 271 (hereinafter "Constraints").

7 William H. Rodgers, Jr., "Choosing a Theory of Liability for Transboundary Pollution," in *Common Boundary/Common Problems: The Environmental Consequences of Energy Production*, Proceedings of a Conference of the American Bar Association and the Canadian Bar Association held at Banff, Alberta, Canada, March 19–21, 1981 (Washington, D.C.: American Bar Association Standing Committee on Environmental Law, 1982), p. 103. (Hereinafter *Common Boundary/Common Problems*.)

8 Armin Rosencranz, "Transboundary Air Pollution: International Law, Agreements, and Other Remedies," in *Common Boundary/Common Problems*, p. 97.

9 "Constraints," p. 257.

10 Nicholas deBelleville Katzenback, "Conflicts on an Unruly Horse: Reciprocal Claims and Tolerances in Interstate and International Law," *Yale Law Journal* 65 (1956): 1096.

11 The minimum contact requirement was created in *International Shoe v. State of Washington*, 326 U.S. 310, 66 S.Ct. 154, 90 LEd. 95 (1945).

12 James M. Fischer, "The Availability of Private Remedies for Acid Rain Damage," *Ecology Law Quarterly* 9 (1981): 437.

13 (1893) AC 602, 63 LJQB 70 (H.L.), rev'g (1892) QB 358 (CA); followed in Canada by *Albert v. Fraser Companies, Ltd.*, 11 M.P.R. 209 (1937), 1 D.L.R. 39 (NBCA), which settled the law despite much criticism.

14 "Constraints," pp. 260–61.

15 Simon Chester, "Remedies in Canadian Courts," *Canada–United States Law Journal* 5 (1982): 86.

16 "Constraints," p. 265.

17 John Swaigen, ed., *Environmental Rights in Canada* (Toronto: Butterworths, 1981), p. 15. It is primarily the provincial rather than the federal attorney general who has authority to act. John P. S. McLaren, "The Common Law Nuisance Actions and the Environmental Battle—Well-Tempered Swords or Broken Reeds?" *Osgoode Hall Law Journal* 10 (1972): 511. In some cases the minister of the environment has authority to intervene. John P. Nelligan, "Remedies in Environmental Law," in *1981 Special Lectures of the Law Society of Upper Canada* (1981): 497.

18   Swaigen, *Environmental Rights*, p. 61.

19   Ibid., p. 92.

20   Chester, "Canadian Courts," p. 402.

21   David Estrin, "Annual Survey of Canadian Law, Part 2: Environmental Law," *Ottawa Law Review* 7 (1975): 402.

22   Stephen C. McCaffrey, "Transboundary Pollution Injuries: Jurisdictional Considerations in Private Litigation between Canada and the United States," *California Western International Law Journal* 3 (1973): 195.

23   Donald Carl Arbitblit, "The Plight of American Citizens Injured by Transboundary River Pollution," *Ecology Law Quarterly* 8 (1979): 345–46. In one other instance U.S. courts became involved with a Canadian polluter. The Ohio attorney general sued corporations from Canada, Delaware, and Michigan over pollution of tributaries of Lake Erie with mercury. The Supreme Court declined to invoke its original jurisdiction and the Ohio courts then issued an injunction against the polluters. This was affirmed by the Ohio Court of Appeals. The case was eventually settled for $700,000. See Richard D. Cudahy, "Clouds on the Horizon: Acid Rain in Domestic Courts," in *Common Boundary/Common Problems*, p. 84. All three courts assumed that the Canadian courts would enforce an injunction issued in the United States.

24   Fischer, "Private Remedies," pp. 433–34.

25   Chester, "Canadian Courts," p. 87.

26   This case is the only one of its kind found after extensive searches of case materials. Its uniqueness is confirmed by many commentators. In addition, the Windsor Utilities Commission, using the same counsel, also sued the *Michie* defendants. *Windsor Utilities Commission* v. *National Steel Corp., Great Lakes U.S. Div.*, no. C-7-72898 (E. D. Mich. 1981) (Gilmore, J.), but this case was dismissed on the grounds that the commission failed to establish a connection between the damage to its ceramic insulators and the pollution. Cudahy, "Clouds," p. 84.

27   495 F2d 213 (6th Cir., 1974).

28   McLaren, "Broken Reeds," p. 509.

29   "First Canadian Suit against Air Polluters in US Courts Ends," *C.E.L. News* (April–May 1975): 40.

30   "Constraints," p. 267.

31   Swaigen, *Environmental Rights*, p. 285.

32   Bruce Mitchell, "The Provincial Domain in Environmental Management and Resource Development," in *Resources and the Environment: Policy Perspectives for Canada*, ed. O. P. Dwivedi (Toronto: McClelland and Stewart, 1980), p. 58.

33   Some of these obstacles may be being removed. In a June 1980 draft policy statement Environment Canada announced it would provide greater access to environmental data. Canada, House of Commons, Committee on Fisheries and Forestry, Subcommittee on Acid Rain, *Still Waters: The Chilling Reality of Acid Rain* (Ottawa: Minister of Supply and Services, 1981), pp. 87–88.

34 Swaigen, *Environmental Rights*, p. 31.

35 Ontario, "A Submission to the United States Environmental Protection Agency Hearings on Interstate Pollution Abatement," Washington, D.C., June 19, 1981, p. 12.

36 O. P. Dwivedi and John E. Carroll, "Issues in Canadian-American Relations," in *Resources and the Environment: Policy Perspectives for Canada*, ed. O. P. Dwivedi (Toronto: McClelland and Stewart, 1980), p. 318.

37 Chester, "Canadian Courts," pp. 85–86. This is true because of the extensive participation of Canadian groups in the U.S. rule-making process, particularly with respect to environmental decisions. Canada would thus be unlikely to deny U.S. groups similar rights.

38 "U.S. Nuclear Plants Threaten Canada, Environmentalists Say," *C.E.L. News* (December 1975): 231.

39 Anthony Scott, "Fisheries, Pollution, and Canadian-American Transnational Relations," in *Canada and the United States: Transnational and Transgovernmental Relations*, ed. Annette Baker Fox, Alfred O. Hero, Jr., and Joseph S. Nye, Jr. (New York: Columbia University Press, 1976), p. 254. In this case the Canadian environmentalists had tacit support from their federal government. David G. LeMarquand and Anthony Scott, "Canada's International Environmental Relations," in *Resources and the Environment: Policy Perspectives for Canada*, ed. O. P. Dwivedi (Toronto: McClelland and Stewart, 1980), p. 95.

40 *Wilderness Society v. Morton*, 463 F.2d 1261 (D.C. Cir. 1972).

41 Wooley, "Current Framework," p. 64. The Department of External Affairs opposes the participation of cabinet ministers and civil servants in U.S. politics. Stephen Clarkson, *Canada and the Reagan Challenge* (Toronto: James Lorimer, 1972), p. 195.

42 Jerome Ostrov, "Interboundary Stationary Source Pollution —Clean Air Act Section 126 and Beyond," *Columbia Journal of Environmental Law* 8 (1982): 73.

43 Ontario, "Submission to the United States Environmental Protection Agency," p. 15. Ontario urged EPA to adopt what it called the "total approach," which involved looking at evidence in the context of a variety of factors rather than in isolation.

44 Wooley, "Current Framework," p. 64.

45 Ontario, Ministry of the Environment, "Presentation to the Michigan Air Pollution Control Commission in Opposition to the Detroit Edison Request to Delay Bringing Its Monroe Power Plant into Compliance with the State of Michigan '1% or Equivalent Sulfur in Fuel' Rule," Monroe, Michigan, June 30, 1982, p. 110.

46 *Wilderness Society v. Morton*, 495 F.2d 1026, 1035 (1974).

47 353 F.Supp. 811 (1973).

48 353 F.Supp. at 818 (1973) (emphasis added).

49 Thomas W. Cross, "Mechanisms for Transboundary Pollution Control," in *Common Boundary/Common Problems*, p. 92 (hereinafter cited as Cross, "Mechanisms").

50 *NRDC v. NRC* 647 F.2d 1345, 1384, n. 139 (1981).
51 "Environmental Effects Abroad of Major Federal Actions, Executive Order 11514, 44," *Federal Register* 44 (1979): 1957ff.
52 627 F.2d 499 (1980).
53 Ibid., pp. 504, 511.
54 Of course, all the usual caveats about the effectiveness of public participation in rule making would apply equally to foreign intervenors.
55 Canada, House of Commons, *Still Waters*, p. 100.
56 Historical information from Van Carson, "The American Legislative Position," *Canada–United States Law Journal* 5 (1982): 75; and "Acid Precipitation: Can the Clean Air Act Handle It?" *Boston College Environmental Affairs Law Review* 9 (1981): 701; Environmental Mediation International, *The Use of Section 115 of the Clean Air Act to Control Long Range Transport of Air Pollution between the United States and Canada* (Washington, D.C., 1981), p. 8.
57 2 U.S.C. sec. 7426(a)(1) (Supp. III 1979).
58 42 U.S.C. 7415 (c).
59 42 U.S.C. sec. 7415(a) (Supp. I 1977).
60 Administrator Costle relied on the IJC's October 1980 *Seventh Annual Report on Great Lakes Water Quality* in his determination that Section 115 had been triggered. "EPA Administrator Believes Canadian Acid Rain Problem May Warrant Action in U.S.," *Environmental News* (January 16, 1981): 1 (press release).
61 Robert E. Lutz, "Managing a Boundless Resource: U.S. Approaches to Transboundary Air Quality Control," *Environmental Law* 11 (1981): 369. The citizen suit provisions of the U.S. Clean Air Act allow citizen suits to enforce provisions of the act and suits against the administrator of EPA to force performance of nondiscretionary acts under certain circumstances. Although these provisions are limited in scope, they do offer at least a partial remedy for citizens affected by some air pollution incidents. It is not clear, however, if these suits are allowed for foreign nationals either under the reciprocity of Section 115 or in general. See "Acid Precipitation," *Boston College Environmental Affairs Law Review* 9 (1981): 740.
62 Environmental Mediation International, *Use of Section 115*, pp. 13–15.
63 For a discussion of the jurisdictional differences, see chapter 6.
64 Environmental Mediation International, *Use of Section 115*, p. 15.
65 A likely alternative source would be a federal-provincial compact.
66 Environmental Mediation International, *Use of Section 115*, p. 18. Citing an October 16, 1978, memorandum to the director of the Office of International Activities at EPA.
67 U.S. Congress, Senate Committee on Environment and Public Works, (*Clean Air Act Oversight Field Hearings*), pt. 6, 1981, p. 42.
68 Wooley, "Current Framework," p. 62.
69 John Roberts, "Transboundary Pollution: Canada's Concerns and Expectations," in *Common Boundary/Common Problems*, pp. 10–14.

70   Carson, "American Legislative Position," p. 74.
71   Information on current status of Section 115 is from a telephone interview with Dolores Gregory, EPA International Affairs Office, Washington,
D.C., March 23, 1983.
72   Gregory S. Wetstone, "Air Pollution Control Laws in North America and
the Problem of Acid Rain and Snow," *Environmental Law Reporter* 10
(1980): 50005. An example of the effect of this focus is the tall-stack
policy, which has increased long-range transport of pollutants.
73   Gregory S. Wetstone, "Legislative Options for Controlling Acid Deposition," in *Acid Rain: A Transjurisdictional Problem in Search of Solution, Proceedings of a Conference*, ed. Peter S. Gold (Buffalo N.Y.:
Canadian-American Center Publications, State University of New York at
Buffalo, 1982), p. 142.
74   Acidic Precipitation in Ontario Study, *Annual Program Report Fiscal
Year 1986/1987*, APIOS Report no. 005, APIOS Coordination Office, Ontario
Ministry of the Environment, July 1987, pp. 39–40.
75   Environmental Mediation International, *Use of Section 115*, p. 23; see
chapter 6.
76   Wetstone, "Air Pollution Control Laws," p. 50012.
77   Nelligan, "Remedies in Environmental Law," pp. 493–94.
78   Ibid., p. 502.
79   Environmental Mediation International, *Use of Section 115*, p. 21.
80   Wetstone, "Air Pollution Control Laws," p. 50014; Quebec does not accept
this power. Jean Piette, "The Role of Provincial Governments in the Field
of Transfrontier Pollution," in *Common Boundary/Common Problems*.
81   Environmental Mediation International, *Use of Section 115*, p. 21.
82   ABA/CBA, "Settlement of International Disputes between Canada and the
U.S.A.," September 20, 1979.
83   Ibid., p. xxxiii.
84   Phone interviews with Dolores Gregory, EPA, and Lisa Lichtenstein, ABA,
March 23, 1983.
85   Phone interview with Lisa Lichtenstein, ABA, Washington, D.C., March
23, 1983.
86   Matthew J. Abrams, *The Canada–United States Interparliamentary
Group* (Ottawa: Parliamentary Centre for Foreign Affairs, 1973), p. 3.
87   La Marquand and Scott, "Canada's International Environmental Relations,"
p. 80.
88   Abrams, *Interparliamentary Group*, p. 9.
89   Dwivedi and Carroll, "Issues in Canadian-American Relations," p. 317.
90   Abrams, *Interparliamentary Group*, p. 120.
91   They covered pollution, national parks, transportation, agriculture, industrial cooperation, inflation, drugs, and China.
92   A separate environmental committee appeared in 1976; the other two
committees were for transborder economic issues and energy. U.S. Congress, Senate Committee on Foreign Relations, *Canada–United States
Interparliamentary Group: Report on the Seventeenth Meeting* (Wash-

ington, D.C.: Government Printing Office, 1976), pp. 2–10.

93   U.S. Congress, Senate Committee on Foreign Relations, *Canada–United States Interparliamentary Group: Report on the Nineteenth Meeting* (Washington, D.C.: Government Printing Office, 1978).

94   Canada, House of Commons, *Still Waters*, p. 95.

95   In another area of interparliamentary cooperation, the Hon. Ron Irwin "participated fully" in a hearing held by the Committee on Environment and Public Works on Oversight of the Clean Air Act on April 14, 1981, even to the extent of questioning witnesses. U.S. Congress, Senate Committee on Environment and Public Works, *Clean Air Act Oversight*, pt. 6.

96   U.S. Congress, Senate Committee on Foreign Relations, *Canada–United States Interparliamentary Group: Report on the Sixteenth Meeting* (Washington, D.C.: Government Printing Office, 1978), p. 8.

97   Clarkson, *Canada and the Reagan Challenge*, p. 195.

98   "Trail Smelter Arbitration," *United States* v. *Canada*, 3 R. Int'l Arb. Awards 1905, 1965 (1941).

99   John E. Read, "The Trail Smelter Dispute," *Canadian Yearbook of International Law* 1 (1963): 213.

100   Ibid., p. 222.

101   "Constraints," p. 254.

102   D. H. Dinwoodie, "The Politics of International Pollution Control: Trail Smelter Case," *International Journal* 27 (1971): 225.

103   "Constraints," p. 25.

104   Hoffman, "State Responsibility," p. 512.

105   Ronald W. Ianni, "International and Private Actions in Transboundary Pollution," *Canadian Yearbook of International Law* 11 (1973): 262.

106   Read, "Trail Smelter Dispute," p. 223.

107   "Constraints," p. 256.

108   Ianni, "International and Private Actions," pp. 264–65.

109   The Canadian victims had been previously compensated by the smelter company and did not face the jurisdictional barriers to lawsuits that the U.S. victims did.

110   U.S. Congress, Senate Committee on Foreign Relations, *The Mexico–United States Interparliamentary Group: A Sixteen Year History*, a report by Senator Mike Mansfield, 1976, pp. 57–58.

111   Senate, *Report on the Sixteenth Meeting*, p. 8.

112   Roger Frank Swanson, *State/Provincial Interaction* (Washington, D.C.: CANUS Research Institute, 1974), p. 83. Prepared for the U.S. Department of State.

113   Richard H. Leach, Donald E. Walker, and Thomas A. Levy, "Province-State Trans-border Relations: A Preliminary Assessment," *Canadian Public Administration* 16 (1973): 422.

114   Maine, for instance, maintains a full-time Canadian affairs adviser, primarily oriented toward commercial issues. Dwivedi and Carroll, "Issues in Canadian-American Relations," p. 309. In 1973 the governor of Maine

and the premier of New Brunswick signed an agreement to maintain cooperation, and protection of the environment headed the list of target areas. See "Joint Agreement between the State of Maine and the Province of New Brunswick," reprinted in "Transnational Regionalism: Energy Management in New England and the Maritimes," *Canada—United States Law Journal* 3 (1980), appendix 3. Maine has also signed agreements with New Brunswick regarding the pollution of the St. John River. Dwivedi and Carroll, "Issues in Canadian-American Relations," p. 315.

115   Maine, Michigan, and Washington are the most active in these. It was one of these conferences that provided the impetus for the Great Lakes Water Quality Agreement of 1972. U. J. Holsti and T. A. Levy, "Bilateral Institutions," in *Canada and the United States: Transnational and Transgovernmental Relations*, ed. A. B. Fox, A. O. Hero, and J. S. Nye (New York, 1976), p. 298, n. 19. Over the past few years the governors and premiers have met in "summits" approximately eighteen times. Roger F. Swanson, *Intergovernmental Perspectives on the Canada-U.S. Relationship* (New York: New York University Press, 1978), pp. 256–57.

116   Maine and Washington State also have legislative exchanges with provinces. Swanson, *State/Provincial Interaction*, pp. 64–65.

117   P. R. Johannson, "British Columbia's Relations with the United States," *Canadian Public Administration* 21 (1978): 218.

118   Swanson, *State/Provincial Interaction*, pp. 11–16. Agreements are defined as having a jointly signed document setting forth regularized interactive procedures; understandings are correspondence, resolutions, and other documents which are not jointly signed but do set out regularized interactive procedures; and arrangements are the remainder of the interactions.

119   Johannson, "British Columbia's Relations," p. 226.

120   Factors based on information in Johannson, ibid., pp. 226–27.

121   John E. Carroll, *Acid Rain: An Issue in Canadian-American Relations* (Washington, D.C.: Canadian-American Committee, 1982), p. 23.

122   Dwivedi and Carroll, "Issues in Canadian-American Relations," pp. 315–16.

123   Johannson, "British Columbia's Relations," p. 231, n. 28.

124   Dwivedi and Carroll, "Issues in Canadian-American Relations," pp. 316–17.

125   Alice Brandeis Popkin, "International Cooperation Begins at Home," *EPA Journal* (June 1978): 10.

126   Clarkson, *Canada and the Reagan Challenge*, p. 196.

127   Ontario, "Submission to the United States Environmental Protection Agency," p. 8; and Norton, "Keynote Address," in *Acid Rain: A Transjurisdictional Problem in Search of Solution, Proceedings of a Conference*, ed. Peter S. Gold (Buffalo N.Y.: Canadian-American Center Publications, State University of New York at Buffalo, 1982), p. 60.

128   Ontario, "Presentation to the Michigan Air Pollution Control Commission," p. 2.

129 International Joint Commission, *First Annual Report on Ontario/ Michigan Air Pollution*, 1976, pp. 21–22.

130 Clarkson, *Canada and the Reagan Challenge*, p. 186.

131 Dwivedi and Carroll, "Issues in Canadian-American Relations," p. 315.

132 Cross, "Mechanisms," p. 89.

133 Gene Albert, "Glacier: Beleaguered Park of 1975," *National Parks and Conservation Magazine* (November 1975): 5.

134 Arbitblit, "Plight of American Citizens," p. 341.

135 Hans J. Peterson, "The Cabin Creek Coal Project: An Environmental Problem and Recommendations," *Social Science Journal* (January 1977): 36.

136 Mont. Rev. Codes. Ann. sections 69-3906(4), -3921.1, -4802(10), and -4820.1(3) (Supp. 1975); and Peterson, "Cabin Creek," p. 33.

137 Environmental Mediation International, *Use of Section 115*, p. 34.

138 U.S. EPA, *Impacts of Airborne Pollutants on Wilderness Areas along the Minnesota-Ontario Border* (Duluth, Minn.: Environmental Research Lab, 1980), p. 2.

139 Environmental Mediation International, *Use of Section 115*, pp. 34–36.

140 Carroll, *Acid Rain, An Issue*, p. 38.

141 EPA, *Impacts*, p. 3.

142 International Joint Commission, *Water Quality in the Poplar River Basin* (Ottawa and Washington, D.C.: Government Printing Office, 1981), pp. 1–2.

143 Arbitblit, "Plight of American Citizens," p. 352.

144 Environmental Mediation International, *Use of Section 115*, p. 32.

145 IJC, *Poplar River Basin*, pp. 8, 57.

146 Environmental Mediation International, *Use of Section 115*, p. 38.

147 IJC, *Poplar River Basin*, p. xi.

148 Johannson, "British Columbia's Relations," p. 224. The Canadian Federal Provincial Relations Office lists more than 250 agreements between the provinces and the federal authorities, and acknowledges that the list is not exhaustive. Kenneth Wiltshire, "Working with Intergovernmental Agreements: The Canadian and Australian Experience," *Canadian Public Administration* 23 (1980): 358. The Federal-Provincial Commission on Air Pollution also exists to provide consultation. International Joint Commission, staff memorandum, "Briefing Notes on Air Pollution for Meeting, July 8, 1980," p. 7.

149 For example, both Washington State and Montana have attempted to restrict the transport of nuclear waste materials through their territory, and Montana has placed a high tax on coal mined within the state that will affect resource allocation decisions elsewhere.

150 Luther C. Heckman, "Banquet Address," in *Acid Rain: A Transjurisdictional Problem in Search of Solution, Proceedings of a Conference*, ed. Peter S. Gold (Buffalo N.Y.: Canadian-American Center Publications, State University of New York at Buffalo, 1982), p. 124. In a single instance the Public Utilities Commission voted on measures involving the expenditure of more than $2 billion for air pollution control.

151 Canada, House of Commons, *Still Waters*, p. 123, appendix 1.
152 Stephen Greene and Thomas Keating, "Domestic Factors and Canada–United States Fisheries Relations," *Canadian Journal of Political Science* 13 (1980): 731.
153 Michael Keating, "Ontario Enters U.S. Pollution Case," *Globe and Mail* (Toronto), March 13, 1981.
154 By "systems" I mean the common-law tradition as opposed to civil law, and by "details" I refer to procedural differences.
155 Involving agencies like the utilities commissions would also bring in something Ontario does have to trade with Ohio—electric power. The power utilities in Canada are public corporations and hence subject to provincial control. Electricity could be sold to Ohio utilities at a subsidized rate in order to ease the burden of abatement costs.

Chapter Ten

1 National Acid Precipitation Assessment Program, *Interim Asessment, The Causes and Effects of Acidic Deposition*, vol. 1, *Executive Summary* (Washington, D.C.: Government Printing Office, 1987), p. I-2.
2 For details, see chapter 3.
3 United States–Canada, *Memorandaum of Intent on Transboundary Air Pollution, Emissions, Costs, and Engineering Assessment, Work Group 3B, Final Report* (Washington, D.C.: Government Printing Office, 1982), p. 1.
4 Canada–United States, *Memorandum of Intent on Transboundary Air Pollution, Report of Legal, Institutional and Drafting Work Group* (Washington, D.C.: Government Printing Office, 1981).
5 These arguments are elaborated in chapter 5.
6 *Congressional Record*, S 8084-8, June 11, 1987.
7 Paul Sabatier and Daniel Mazmanian, "The Implementation of Public Policy: A Framework of Analysis," *Policy Studies Journal* 8 (1980): 555.
8 See chapter 9.
9 National Acid Precipitation Assessment Program, *Executive Summary*, p. I-10.

||||||||

||||||||

Contributors

||||||||

Project Directors

Dr. Jurgen Schmandt, Professor of Public Affairs, University of Texas at Austin. Formerly senior environmental fellow, U.S. Environmental Protection Agency; associate director, Harvard University Program on Technology and Society; head of national science policies unit, Organization for Economic Cooperation and Development, Paris.

Dr. Hilliard Roderick, Distinguished Visiting Tom Slick Professor of World Peace, University of Texas at Austin (1982–83). Formerly director of enviromental affairs, Organization for Economic Cooperation and Development, Paris; former deputy director, Fusion Research Program, U.S. Atomic Energy Commission.

Contributor to Second Edition

Dr. Judith Clarkson, Environmental policy analyst and free-lance writer. Formerly associate biologist, University of Texas System Cancer Center.

Graduate Students

Tom Albin, B.A., M.A. English, Louisiana State University.
Barbara R. Britton, B.S. Nutrition, University of Texas at Austin.
Kim J. DeRidder, B.A. Interdisciplinary Studies, University of Tennessee.
Mary Anne Doyle, B.A. Political Science, Tulane University.
Robert Egel, B.S. Conservation of Natural Resources, University of California at Berkeley.
Richard Henderson, B.A. Astronomy, University of Texas at Austin.
Pierce Homer, B.A. Philosophy, Haverford College.
John Kane, B.A. Political Science, University of Texas at El Paso (deceased).
Paul Kinscherff, B.S. Public Affairs, University of Southern California.
Mike Mgebroff, B.A. Political Science/History, Texas Lutheran College.

Kimberley Mickelson, B.A. Plan II Honors Program, University of Texas at Austin.

Andrew Morriss, A.B. Public Affairs, Princeton University.

Steve Paulson, B.A. History, University of Iowa.

Robert Stewart, B.A. Government, University of Texas at Austin.

Katherine Wilshusen, B.A. Sociology, Denison University.

Index

tion of, 117–18

National Coal Association (NCA), 119; detailed position of, 123–24

National Commission on Air Quality: finds air quality improvement, 160; recommendations of, 165–66

National Conference of Commissioners on Uniform State Laws, 232

National Emissions Guidelines, 140–41

National Emissions Standards, 141

National Energy Policy (Canada), 79

National Environmental Policy Act, 225

National Governor's Association, 172

National Governor's Conference, 100

National Wildlife Federation, 118, 135

New Brunswick, 142; and Maine, 242

New Clean Air Act, 175–76

New Hampshire, 177, 260

New Source Performance Standards (NSPS), 162, 166, 172–73

New York, 177, 230; and Ontario intervention, 248

Nierenberg, Bill, 92

Nitrogen, 49; compared to SO_2, 109–10; oxides of, 8–10, 15; principal sources of, 40, 109–10, 138; trends in production of, 39, 110

Nixon, President Richard, 198; calls for controls, 160; reverses stand on phosphates, 199–200; signs 1972 agreement, 201

Noranda mines, 155

North America Air Quality Commission, 235

Norton, Keith, 153

Nova Scotia, 137

Nuclear power, 18, 154; and Ontario Hydro, 130, 133–34

Office of Environmental Affairs, 199

Office of Management and Budget, 29, 169

Office of Technology Assessment (OTA), 111, 112

Office of Science and Technology Policy, 63, 90, 100

Ohio, 229–30, 260

Ohio Public Utilities Commission, 247

Oil-burning sources, 123

One-nation approach to acid rain dispute, 249

ON-HY. See Ontario Hydro

Ontario, 142, 148–57; Acidic Precipitation in Ontario Study (APIOS), 152–53; Environmental Protection Act, 148, 151; and INCO, 131–32; intervenes in U.S. regulation, 224, 230, 248; Ministry of the Environment, 149, 156, 207; and Michigan-Ontario Transboundary Air Pollution Committee, 244; presses for U.S. action, 155–57, 243; regulation of SO_2 sources in, 151–52, 153–54; research and monitoring in, 152–53

Ontario-Canada Task Force, 145

Ontario Hydro: and Atikokan Power Project, 244–45; and Canada Coalition on Acid Rain, 130; detailed position of, 133–34; and nuclear power, 130, 133; regulation of, 153–54

Organization of Science Counselors, 100

Ottawa statement on development of acid rain agreement, 65–66

Ozone, 43, 110, 112

Parker, Larry, 16, 135

Parliamentary Center for Foreign Affairs and Trade, 234

Peabody Coal Company, 123

Peer review of work groups investigating acid rain, 90–92

Perley, Michael, 130

Pennsylvania, 230

People of Enewetak v. Laird, 225

Phosphorus, 198–201

Policies for controlling acid rain: